Ku

Kuliba

Rider of the Pale Horse

Rider of the Pale Horse

A MEMOIR
OF LOS ALAMOS
AND BEYOND

McAllister Hull

with Amy Bianco

Illustrated by John Hull

University of New Mexico Press

Albuquerque

For Mary, who lived through it all

©2005 by the University of New Mexico Press
All rights reserved. Published 2005
Printed in the United States of America
10 09 08 07 06 05   1 2 3 4 5 6

LIBRARY OF CONGRESS CATALOGING-IN-PUBLICATION DATA

Hull, McAllister H., 1923–
  Rider of the pale horse : a memoir of Los Alamos and beyond / McAllister Hull
with Amy Bianco ; illustrated by John Hull.
        p. cm.
  Includes bibliographical references and index.
  ISBN 0-8263-3553-5 (alk. paper)
        1. Hull, McAllister H., 1923–
        2. Physicists—New Mexico—Biography.
        3. Los Alamos Scientific Laboratory.
        4. Manhattan Project (U.S.)
        5. Nuclear weapons—Social aspects—United States—History.
        6. Cold War—Social aspects—New Mexico—Los Alamos.
        7. Nuclear energy—Research—United States—History.
    I. Bianco, Amy, 1964–   II. Title.
QC16.H85A2 2005
    530.092—dc22
                        2005011952

DESIGN AND COMPOSITION: *Mina Yamashita*

Contents

Preface

The first version of much of this memoir was written in late 1945 or early 1946 after I transferred to T-division at Los Alamos National Laboratory and began to work on the Bikini Test phenomenology. I wanted to see whether I had acquired a reasonably complete understanding of the nuclear physics and technology of the two nuclear bombs we had built for the war. I now had access to any document I wanted to read, but not as much leisure as I would have liked, since we were still working with some of the wartime urgency. So I just sat down and wrote out the story. It was fairly nontechnical, partly because I was still learning nuclear physics—a process that is still going on. I left the written piece on my desk in the Gamma Building and found the next day that it had been picked up by security and classified. I could have gotten it back, but I didn't bother, since I had written it only for my own clarification. I suppose it's still in the files.

When I began to teach at Yale in 1951, I always included a lecture or two on nuclear energy and nuclear weapons and discussed—with debate if possible—the attendant policies I thought needed to be considered for a sound national agenda on nuclear matters. This memoir is, for me, an extension of that practice for a wider audience. I have strong—and, I trust, well informed—opinions on the employment of nuclear reactors for the energy needs of the world and the development of nuclear weapons under any conditions. I try to leaven my argument here, especially on weapons, by some personal history that tells how I got where I am.

I didn't write anything for anyone but me to read until recently.

My oldest granddaughter, Damaris, asked for the story of nuclear weapons ten or twelve years ago, and so I wrote something for her. Very little of what I knew was still classified by then, so the story was fairly comprehensive. She later had me invited to talk to her history class in high school. Shortly afterward, my son, John, asked for a narrative he could use to inspire a suite of paintings on the S-Site work. Fortunately, we were able to visit the main casting building at Los Alamos before it was derelict and torn down. Some of these paintings were in an exhibition in the Governor's Gallery in Santa Fe to help raise private funds to preserve some of the wartime buildings at Los Alamos. John donated one of the paintings to the permanent collection of the Bradbury Museum at Los Alamos. He has seen the book in various drafts as he created completely new illustrations. It's the first time we've worked together this way.

My classes in Physics and Society that I started in Buffalo and brought to some kind of good order at the University of New Mexico again included a section on nuclear energy and nuclear weapons. I kept promising to write out a set of organized materials for the students because the course was different enough that available texts were not very helpful. I finally wrote the nuclear physics part a year or two before I retired for the second time and gave up teaching. I planned to get this part published, and if possible, write for the other parts of the course, but the press of my university doesn't do texts, so I had to go elsewhere. I tried one of the big New York publishers, and an editor there, Amy Bianco, thought the chapter in the material on weapons could make a trade book — if I was willing to tell a personal story. This book is the result of the collaboration between me and Amy: I wrote in my usual linear, didactic style, and she rearranged the written material into a narrative she — and the University of New Mexico Press (which *does* publish memoirs) — hoped is more attractive to intelligent readers who are as interested in the human story as the technical one. Working with a collaborator is a new experience for me, as is writing in the first person. Amy asked for some personal anecdote she wanted to include and demanded fuller explanations of technical matters when I was not clear. We have worked well together, and the results are much better than if I'd worked alone. I remain interested in bringing an issue that is critical for the future of our world before the readers who can determine the policies of their governments. Of course, I hope the readers will also find this a good story.

The work I describe at S-Site was not done alone, so I wish to remind readers of the hard work of the few GIs who helped us learn to cast reliable lenses and the larger number of civilian powder men who cast them in production for the Trinity Test (now sixty years ago) and Nagasaki.

Some of the material appearing here has been published before, and we are grateful to the World Scientific Publishing Company for permission to reprint some of what I wrote for the Gregory Breit Centennial Symposium proceedings that they published, and to the National Academy of Sciences and its publisher, the National Academy Press, for permission to reprint some of the memoir I wrote on Gregory Breit for their member biographies series.

We thank Phill Bloedow, Department of Chemistry, University of Wisconsin, for finding us a picture of Joe Hirschfelder taken in about 1947 or 1948, just after he returned to campus from Los Alamos. William R. Massa, Public Services Archivist at Yale, took a new picture of Sloane Physics Laboratory for us, and Bob Gluckstern found a picture of himself from those ancient times when we worked together.

The staff at the press does the hard work, and we thank Beth Hadas, who is supposed to be partially retired, for taking on the book.

—McAllister Hull
Albuquerque
September, 2004

**Behold a pale horse: and his name
that sat on him was Death**

—Revelation, 6:8

Introduction

When I was ten or eleven years old, some school assignment or item of curiosity sent me to the modest encyclopedia in our family library to look up the article on light. The piece was not very long, as I recall, but it included a sentence that has influenced my life for the last seventy years or so. The sentence read: "According to the theory of relativity, light has weight."

Now, Steven Mithen and Steven Pinker, among other students of the human mind, believe we are born with a built-in understanding of physics—enough to succeed as hunter-gatherers, in any case. Whether that innate understanding extends to light and weight I doubt, but as a boy I immediately understood that the scale on which I weighed myself was not going to work with light. The statement in the encyclopedia was not developed, and my books said nothing about any physics, much less relativity. So I went to our local (Waco, Texas) Carnegie Library, where I was a regular customer for boys' adventure stories and books on astronomy, the only popular science I had ever found there. The librarian, starting with "relativity," came up with three items of interest:

(1) Albert Einstein was responsible for the theory.
(2) He was called a "physicist."
(3) A book called *Easy Lessons in Einstein* by a man named Edwin Slosson.

I quickly read the book—it was very slim—and learned about shrinking meter sticks, slowing clocks, light rays bending in the gravitational field of the sun, and the perihelion of Mercury. Obviously I got more than I started out to discover, which is the nature of education—or ought to be. My fascination with these ideas and my newfound awareness that people who study these matters are called physicists determined my career choice. I decided to become a physicist.

I knew I wanted to be a physicist, but I didn't know how a physicist made a living, and I could never have anticipated the course my career would take. I began my formal studies at Mississippi State in 1941 in an engineering program, thinking that engineering was something I could teach while I worked on physics problems. We entered World War II in December of my freshman year, so I took a job as a draftsman in an ordnance plant in the summer of 1942. What I learned there, working with problems of explosives in the chemistry department, would have more of an impact on the course of my life than any of my academic training to that time, for it would take me to Los Alamos. I was drafted in March 1943, and in the fall of 1944 I was sent to Los Alamos to cast explosives. Though I would spend only about two years at Los Alamos itself, I would be involved with that unique place in one way or another for the rest of my career.

After the war was over, I helped calculate the phenomena attendant to the explosion of nuclear weapons in the Bikini Tests. During this time I did the research for my first published paper on the penetration of gamma rays through thick targets, which involved quantum electrodynamics. I was between my sophomore and junior years of college and had not yet taken a proper course in classical mechanics, much less quantum mechanics. After I was discharged, I continued to work at Los Alamos until school started in the fall of 1946. Los Alamos was to shape my life even further, for I met my major professor there in 1946 and moved with him to Yale in the middle of my junior year. Gregory Breit and I worked together for twenty-five years in New Haven, occasionally on special problems for Los Alamos. In 1976 I returned to New Mexico as provost of the University of New Mexico (from which post I am now retired), and I continue to have interactions with the Laboratory.

As I studied physics in a somewhat more orderly fashion in courses at the university, the central ideas of the discipline became clear to me. I learned how they were integrated into a picture of the physical universe and how they led to new concepts. Physics is the search for the fundamental laws that govern the functioning and structure of the universe and all the physical objects and processes in it. Physicists thus undertake a daunting task that may never be completed. They are inspired by two fundamental assumptions: (a) that the world they study exists outside their minds, and, when they have gotten its properties right, it will be seen in the same way by other physicists; and (b) that in the process of getting it right they must search deeply enough beneath the surface of what they observe so that no more fundamental level can be found. The one-word labels for these beliefs are Platonism and reductionism—so physicists are, in general, Platonists and reductionists.

The average working physicist does not think of these labels as describing his worldview, and he certainly does not worry about entertaining concepts that may not strictly conform to philosophers' definitions. Physicists look for consistency among the concepts they adopt; any given concept must fit with all others that have been defined in the field, and the phenomenon it describes must be the same wherever and whenever it is encountered. Above all, physicists want the referents of their concepts to remain unchanged as processes involving them continue to develop.

Physicists do not believe (yet?) that they can describe the total reality of the world, only parts of it. We know that the reductionist approach does not cover all the phenomena we try to explain. (I once complained to a biologist friend that his systems were too complicated to be understood from the atom up. The advent of complexity theory and the study of self-organizing systems only justify my complaint!) Our theories are models of the ultimate objective physical reality we believe we are studying, and, as models, there are limits to their faithfulness of representation. Of course, we do not agonize over this all the time. We are pragmatists and think in terms of what works within the accepted epistemology of the field.

Heraclitus reminds us that we cannot step in the same river twice, for change is the way of the world. But there is a constant: the river remains. Constants in the midst of change are very important to us, as they have been to people who think about the nature of the world since, perhaps, Greek ideas began to focus in the middle of the first millennium BCE. Matter was an early concept of something that persisted, and in the sophisticated ideas of Democritus (expounded by Lucretius), matter is composed of atoms, which accounts for the apparent loss of matter when a log burns. Some of the atoms comprise the smoke that rises from the burning log. Note that reductionism has triumphed over observation here. The atoms cannot be seen, but they yield a description a level below the obvious description of the log as a wooden cylinder, say two feet long and six inches in diameter. These unobserved atoms provided these early thinkers with a concept that (a) allowed the conservation of matter to be retained, since atoms don't change; and, (b) allowed consistent descriptions of different kinds of matter on the same terms, since it held that they were composed of different compounds of the few, indestructible atoms.

The evolution of the concept of "atom" provides a good illustration of the process of learning about physical reality. In the 2,500 years since the atom was first postulated, its meaning has undergone development and extension, but its role as the basic unit of matter has remained. The atom of the Greeks was indestructible, and so it remained as chemists, especially, parsed the four classical elements of fire, air, water, and earth into the ninety-two modern elements of the periodic table—plus two dozen more we have made in the laboratory since 1940—each identified by its characteristic atom. But twentieth-century physicists found that the atom, while retaining its position as the elementary constituent of matter in the myriad compounds that make up our world, was not indestructible after all. It had parts and a structure: a cloud of electrons surrounding a nucleus of neutrons and protons. The electromagnetic field holds the electrons in orbit, and the strong force holds the nucleus together. The electrons can be separated from the nucleus, and the nucleus itself can be resolved into its constituents. Current ideas take us two levels below this picture, where nucleons (neutrons and protons, the constituents of the nucleus) are made of quarks, gluons carry the strong force, and quarks, gluons, and electrons are made of superstrings.

It is not known whether we have now arrived at the final, basic level of explanation. There has been talk for several years of a "Final Theory" or a "Theory of Everything" that would unite the forces of the atom with all other known forces, but physicists know better. Even if we find such a theory, it will, at best, be nothing more than a theory of the physical universe—a very important part of what may be conceived as the "whole," but certainly not all of it. In any case, physicists—as pragmatists—work at whatever level of explanation is needed to address the problem at hand. There is no need, for example, to invoke superstrings to discuss nuclear physics. Fission and the release of nuclear energy may be treated almost classically.

It is apparent, then, that the knowledge of the physical world we develop evolves and is, at any moment, contingent. Newton's theories of motion and of gravity could, together, predict the orbits of the planets and even reveal the existence of planets not yet known at the time (Neptune and Pluto), but it could not explain the details of the orbit of Mercury. This is what we mean when we say theories are contingent: they are valid over a wide range of values of their variables (position, mass, velocity, time, and so on), but not all values. Newton's theories remain valid (if incomplete) descriptions of the world for the range of variables for which they were formulated in the first place—velocities small compared with the velocity of light and masses small compared with the sun's mass. We design space vehicles and launch astronauts according to Newton's theories, but we explain the advance of the perihelion (the point in the orbit closest to the sun) of Mercury with Einstein's general theory of relativity.

Einstein's method of developing general relativity is instructive here. It was well known before Newton's time that the measure of resistance to change of motion, called inertia, is equal to the parameter that gives the strength of gravity, although Newton was the first to give this equivalence a clear and productive formulation. (Whether or not Galileo actually did demonstrate the equality of inertial and gravitational mass at the Leaning Tower of Pisa, he certainly knew the result.) With Einstein this became the principle of equivalence. From his own theory of special relativity, Einstein also knew that mass and energy are equivalent, and that this result must come out of any general theory of relativity. In addition, Einstein opened a new approach to constructing physical theories by insisting that they be independent of any local transformation of coordinates. His reasoning was that since the systems we choose to describe the world are arbitrary, their choice cannot affect the physics that occurs. In special relativity this means that physics must be the same in all inertial—or nonaccelerating—systems. Obviously, inertial systems are a subset of all systems, so in the regime of validity of special relativity, any general theory would have to give the same picture of physical reality. Einstein further insisted that his description of gravity yield the same results as Newton's theory for appropriate masses and velocities. When he finally got the formulation he wanted, he found that gravity was replaced by curved space-time, so an accelerated frame and one for which there was a gravitational field were equivalent—an observer could not tell the difference from inside the frame. He found that the current best value for the anomalous advance of the perihelion of Mercury was given exactly by his theory. He said afterward that this result was the most satisfying in his life. Having made a few (much less momentous) calculations myself, I think I know how he felt. Einstein's intuitive approach in developing general relativity—starting with the symmetry one wishes to achieve rather than looking for a symmetry to fit data—is the approach favored today in looking for theories that embrace all the forces we know about in the physical world. (His attempt, though classical, to unify gravity and electromagnetism was in the same spirit, but two new forces, the weak and the strong nuclear forces, were discovered while he labored, so his approach to that problem was hopeless.)

The lesson here is that a theory that has been thoroughly vetted by experiment is always useful in the regime where it has been confirmed, as is Newton's theory of gravity. In forming new theories to account for unexplained facts, it is helpful to start from older, validated formulations. This continuity in the development of physics is why most of us find Thomas Kuhn's idea that science advances in abrupt steps called paradigm shifts inappropriate. The facts he deals with are valid enough, but they are not the whole story; and he fails to understand the way physics really develops. Of course, Einstein's curved space-time is different from Newton's gravitational fields, but the latter are contained in the former, and they transform smoothly into each other at their boundaries. They are models of the same reality. It is interesting that today, almost eighty years after Einstein formulated this theory, general relativistic corrections are necessary to allow Global Positioning Satellites to fix positions on earth to the precision required by the users. Thus a theory formulated to treat the universe in the large is applied to navigating cars in a tiny corner of it! To one who has followed the difficult search for experimental verification of general relativity beyond light bending and Mercury wobbling, this mundane application is fascinating.

Newton's equations are inadequate for the very small as well. The development of quantum mechanics, the name we give the theory of the very small, took yet another path from established physics—and at about the same time. Following Max Planck's introduction of quanta of radiant energy (a quantum is just a piece of something; for example, a penny is a quantum of money), Niels Bohr applied the idea to the orbits of the electrons that attend nuclei in Ernest Rutherford's "solar system" atoms. He began with Newtonian orbits, but insisted that only some of them were realized in nature. He quantized them. Other formulations of classical mechanics were used to develop more broadly applicable quantum theories. Paul Dirac incorporated special relativity, and found that Wolfgang Pauli's spins for the electron came out naturally. Antiparticles were predicted and eventually found. When quantum mechanics is applied to electrodynamics, the best-validated theory in physics is formed: quantum electrodynamics gets agreement with measurement to twelve decimals—hardly an accident! It is an interesting comment on the pragmatism of physics that the wave functions of quantum mechanics are not observable in principle—just their absolute squares. A completely acceptable interpretation of quantum mechanics—especially for measurements—has eluded us to this day, yet we calculate quantities that are measurable with great success. Superstrings are as yet still in the hypothesis stage, but their formulation uses ideas from classical mechanics. The continuity of our models of physical reality thus extends from the edge of the universe to the constituents of nucleons, a span of some forty orders of magnitude (or perhaps twenty orders more, depending on the string model we use). That range is startling even to those of us engaged in current speculations about how the world works.

Part of the power of physics in describing the physical world lies in its formulation in terms of mathematics. Galileo started this approach, and it has continued ever since. (The Greeks had used mathematics, but it was not a requirement for their models.) It is puzzling to practicing physicists that this is so. The efficacy of mathematics in physics is unreasonable, suggested Eugene Wigner, one of the giants in twentieth-century theory. When Einstein needed a way to formulate his general relativity, Riemannian geometry* was waiting. Hilbert space† was around before quantum mechanics came along, and Lie‡ had described his groups before they were needed for particle theory. (However, when Newton needed a means of working with his formulation of gravity, he had to invent the calculus!) Despite its beginnings in observations of the world (think of the string stretchers of Egypt, for example, or the stone counters of Europe), mathematics is a free creation of the human mind, while physics remains essentially attached to observable reality. Perhaps the answer to Wigner's "unreasonable efficacy" lies in the fact that the same kind of mind constructs the objects of mathematics and chooses the concepts of physics.

The discipline of physics requires that the validity of even the most far-flung concept or theory must be tested in circumstances other than those in which it arose—another important source of its power. Ideally, it should be possible with a well-formed theory to predict in detail the outcome of such experiments; agreement between prediction and experiment provides confidence in the "truth" of the theory or the validity of the concept as a description of the objective reality we are attempting to understand. The experiment itself is an unassailable

---

*Bernhard Riemann was a mathematician of the nineteenth century who developed the theory of curved spaces (differential geometry). This was just what Einstein needed to describe curved space-time in general relativity. I tried to teach myself differential geometry (and tensor analysis) in high school. I also read science fiction, which I'd come across at about the same time as general relativity: stories of Isaac Asimov, John Campbell, E. E. Smith, Ray Bradbury—easier reading than differential geometry!

†David Hilbert was a mathematician who studied abstract vector spaces, where the vectors are functions with suitable definitions and properties. Solutions of Schrödinger's equation are vectors in a Hilbert space and hence may be studied by Hilbert's methods.

‡Sophus Lie studied special groups that find broad uses in mathematics, but are especially appropriate for relating the properties of families of elementary particles.

truth—within the limitations of its measurements. But as a candidate for absolute truth . . . there's the rub. No measurement is infinitely precise, and a a better approximation to the reality we believe is there may lie within the experimental uncertainties (due, for example, to limitations in the apparatus). Obviously this means the measurement is not absolute, nor is the theory that predicted the result of the experiment. Even quantum electrodynamics contains quantities that cannot be calculated exactly within the theory. Approximations must be made, so we compare a theoretical result of limited validity with an experimental result with technical limitations! But both are valid within the uncertainties of the measurement and the theoretical approximation, and the agreements are breathtaking.

Physics does not provide absolute truth (nor does mathematics, as Bertrand Russell and Kurt Gödel have shown us), but its theoretical constructs embrace an ever greater range of phenomena in a broadly consistent way, and, except at the outer limits of our search into the nature of physical reality, physicists agree on the meaning of concepts and the nature of theories. As I grew up in the field, in a somewhat helter-skelter way that in the earliest of my studies paid little attention to the usual hierarchical presentation of physics in school or university, it was the power of well-formed theories that always held my interest. I never thought the knowledge developed by physics was certain in an absolute sense, but I found it offers reliable models that allow us to understand physical reality in a fundamental and repeatable way.

Though we cannot know what we know with absolute certainty, we can have sufficient confidence in our theories to put them to practical uses. Our technological society is built on the applications of physics, and the abundant amenities of our society come from all the sciences. Because our theories are rooted in observation of the physical world, are formulated in mathematics that allows solutions to unanticipated practical problems to be found, and are reductionist and continuous between levels (so that only the level of explanation required for a particular

application need be employed), technology can mine the basic knowledge of physics with confidence in making its devices. The laser needs
quantum electrodynamics, and the computer chip relies on solid-state
quantum mechanics. Light meters and door openers use the photoelectric effect, a quantum phenomenon explained by Einstein, for which he
got the Nobel Prize. (Relativity was too controversial for the committee!)
The pervasive electrical power industry operates with classical electrodynamics. The communications industry also depends on James Clerk
Maxwell's (classical) electrodynamics; the waves he found now carry
radio, TV, and cell phone signals. We launch communications satellites
by Newton's classical mechanics and operate them with classical electrodynamics. We put astronauts in space with classical mechanics and
Newtonian gravitation. (Rockets operate on Newton's third law of motion,
and he anticipated low earth orbits!) We need relativistic quantum

mechanics to understand the nucleus fundamentally, but we can make nuclear reactors and bombs with little more than classical ideas.

Physicists ordinarily do not work on these applications; they are interested in going on to the next problem in their study of physical reality. The distinction between physics and engineering is not hard and fast, however, and it can take years for basic findings to yield useful technologies. Heinrich Hertz observed Maxwell's waves twenty-four years after their prediction, and Guglielmo Marconi sent messages across the Atlantic thirteen years after that in 1901. It took nearly another twenty years and its use in WWI for radio to become a practical technology for general use. (Anyone who has tried a crystal receiver for radio reception may question how "practical" early radio really was.) Thus in this case the application followed the basic discovery by nearly sixty years, and physicists made the first two steps in reaching that stage.

In wartime this steady and deliberate process of finding useful applications of physical discoveries always appears too slow. Consequently, physicists were directly involved during WWII in the U.S. and Britain in the development of a number of technologies crucial to the war effort, such as the proximity fuse, which allows planes to be downed with a near miss; radar, which allows them to be seen at a distance (the Battle of Britain was won at least partially because of British radar); the means of making steel ships less susceptible to magnetic mines; and nuclear reactors and weapons. In Germany physicists worked on long range ballistic and cruise missiles. In the Allied nuclear program the time between the discovery of fission and the first operating reactor was just six years—perhaps only four counting from when we knew what the first experiments meant. Enrico Fermi was involved in both and saw the effort through another three short years to the test of the first nuclear (so called "atomic") bomb. For good or evil, we had shortened the time between discovery and practical application dramatically.

The idea of physicists (or chemists or biologists) working on weapons worries those of us who have done it. Engineers and technicians are focused on accomplishing a specific, set task. It is not in their culture to consider the larger implications of their work. Neither do physicists generally look around much or contemplate the possible consequences of what they do, in part because the applications come years later. But because those of us who worked on the first nuclear weapons did the development and engineering as well as the basic research, we were well aware of the consequences all along. As J. Robert Oppenheimer once said: "For the first time in history, physicists have known sin"—a little dramatic, but he had a point.

The technology that underpins our society comes from the unfinished, non-absolute, and contingent physical theories we have. Some technologies might harm the environment, or even put the survival of the human race at risk. Their cost and benefits need to be determined ahead of their deployment, at least to the level of assurance that John Stuart Mill tells us is all we can expect. As unrealistic as this idea may be in practice, it must be considered, and it is a responsibility of those of us with expert knowledge to so inform the decisions of our society. Obviously, war modifies the criteria we may use to justify the creation and deployment of any given technology, and the defense of our values will modify them in peacetime. An example is the peacetime development of the hydrogen bomb, which cannot be treated as just another weapon in the arsenal of the military, but at best as a tool of deterrence. Even if deterrence were to fail, it would be very difficult to use the hydrogen bomb rationally because it is so devastating.

This caution is in no way to be interpreted as a suggestion that any topic in basic research should be curtailed. Basic research in all fields provides the intellectual capital from which we can draw as need (often unanticipated) arises. It is a wise government that supports such research, for—if for no other reason—it must stay ahead of any present or future enemies.

It is the selective further development, deployment, and application of basic knowledge that must be a decision of society as a whole, not of individual scientists, engineers, or entrepreneurs. The physicists who make the basic discoveries and the engineers who exploit them can claim expert knowledge of how a device or process works. We can gauge better than most what the consequences of deployment of one of them will be, and in some cases, we can even forecast a technology that hasn't yet been developed. But policy is a matter for all of us.

We have no final theory of physical reality, much less one for how life arises and the mind works. But reason is an observed attribute of the human mind, and if it cannot handle all the problems we humans make for ourselves, it should still be pursued as far as it can take us. In addition, as Thomas Jefferson has said, the members of the deciding society must be informed if they are to make optimum decisions. The drastic impact of nuclear weapons has induced many of us in the scientific community to try to inform the public about the dangers of their application in a general war. And science is always raising new questions. Whether to allow cloning of humans is a recent one. Some of us write about such issues for general consumption; some of us construct courses for college students in the hope that they will carry away not only some useful information, but also some methods of thinking about new questions as they arise. Rational analysis and creative synthesis—the beginnings of physics epistemology that I have outlined here—will not produce absolute truth about anything. But the practice, rare enough and sometimes quite lonely in our society, of recognizing one's premises and reasoning from them as far as the limitations permit can reduce the uncertainties in dealing with a dynamic world to the vestigial minimum inherent in human interactions. It surely worked for me during the war and afterward, especially in the difficult task of trying, if not entirely successfully, to lead a major university to realize its potential.

one

S-Site

The War Effort

On my sixteenth birthday Hitler invaded Poland. It was September 1, 1939, and this act of aggression set the course of my life. In December of 1941, when the U.S. entered the war, I was eighteen and healthy. I assumed I was draft bait. I didn't realize that as a member of the Coast Artillery ROTC at Mississippi State, where I was a freshman, I could probably have finished my course in electrical engineering on a hurry-up schedule and entered active service with a bar. As it turned out, I was to be the first of my family to finish college. My father's brief college career, following his own army service in WWI, where he lost three fingers of his left hand, had ended with his marriage to my mother. So he had no advice to give me. He thought I'd probably be drafted within a few months.

I got a job during the summer of 1942, hoping to be useful while I waited for the draft. It didn't occur to me to enlist. The job was with the Midland Ordnance Foundation in Illiopolis, Illinois, a small town a little nearer Decatur than Springfield. The Foundation hired me as a draftsman and tool designer. I'd had a year of drafting in college and no foggy notion of tool design, but I figured I could learn.

My first weeks on the job were spent drafting organization charts for Midland Ordnance, which had just been set up as a new company by Johnson and Johnson for the war effort. The parent company supplied management talent, and the Ordnance Department explained what needed to be done to make artillery shells. This kind of redirection of effort was common in those days; in less than a year, Johnson and Johnson had turned from making baby powder to making Bofors 40-mm antiaircraft shells. Any heavy manufacturing company could make the big guns that served the artillery. Automobile makers turned to tanks. Airplane manufacturers increased production and new designs tremendously. Gun manufacturers all made the rifles we needed, ignoring patents, as did the makers of army trucks and the famed jeep. Women joined the workforce, never to be relegated again solely to homemaking, and Rockwell's "Rosie the Riveter" became an icon. This universal war effort was why the Japanese Admiral Yamamoto told his Emperor that he could win against the Americans for a year, but after that the U.S.'s great industrial capacity would prevail. The effort in Illiopolis was a small part of making his forecast come true.

The first thing I was asked to design was a table for the production line on which the machines used in making Bofors 40-mm shells were to sit. I dutifully looked into the dimensions of the machines and where the operators would stand or sit and produced a design. When the shop made one, I was horrified; it was the most graceless piece of furniture I had ever seen. It haunted me for the rest of my stay in Illiopolis because I saw dozens of them on the plant floor whenever I went in there. It turned out that the tables suited the job quite well!

Soon I was transferred to the chemistry department (I had had a year of general chemistry, too), where we did quality analyses on the explosives that were used to fill the shells the company was making. So I taught myself about explosives—their chemical formulas, densities, melting points, and so on—and what would set them off! The formal study of explosives was limited then to understanding the chemical changes that took place when an explosive was ignited. Most explosives had been discovered by powder companies using trial and error—most famously, perhaps, that of Alfred Nobel, who found that mixing nitroglycerin with an inert carrier made a reasonably safe explosive he called "dynamite." Thus, while the empirical study of explosives—burn rates especially—was as old as Greek fire and black powder, the theoretical science of explosives was in its infancy. But I learned what I could from my books.

It became my job to introduce new members of the department to the handling of the explosives. To reduce their natural fear of these materials, I would usually hit a piece against the concrete floor with a hammer. (Of course, I carefully selected the explosive and the hammer stroke for this demonstration.) Production involved machining cast explosives (usually TNT) inside a shell to hold the fuse, so one had to know what was possible without danger. The machine tools were made of monel metal, a copper amalgam that would not spark when it struck steel.

The chief chemist, who was not an explosives expert, gave me considerable responsibility in the lab. One difficult problem we had concerned the tracer for the shells. The shells we were making were for antiaircraft guns, and it was helpful for those who would be shooting them to be able to tell where they were putting their ammunition as they fired at an attacking plane. So we added a chemical tracer to the shells. When the round was fired, the propellant caused the tracer to burn with a bright glow that could be seen easily. When it burned through, it ignited the TNT charge. The tracer was a mixture of compounds that included organic molecules. The operators mixed up the tracer, which had the consistency of brown sugar, and then forced it into a hole in the back of the shell. The problem had to do with the mixture: if the formula was wrong or the mixture was not homogeneous, the tracer malfunctioned. I tried to design foolproof procedures, but they didn't always work in the rush of a moving production line. We sampled the mixtures, but the analysis by standard chemical methods took so long we couldn't risk leaving the mixture to separate while we checked it. So we allowed the mixture to be put in the shells and held the batch until the analysis was done. If the analysis failed, then we had a batch of useless shells to store.

One day the chief chemist mentioned to me that he knew of an analysis procedure called "polarography" that might be faster, but that he didn't know much about it. Once again, I got a book and taught myself what I needed to know. Polarography is a chemistry technique that uses physical processes, so it fit my interests well. I got the equipment and began to experiment. In a few weeks, I had worked out an analysis method for our tracer that agreed reasonably well with the traditional method and took only an hour. Thus the analysis could be done before the shells were filled, and we had no more spoiled batches.

Army Training

In due course the spring of 1943 appeared, and with it came my long-expected draft notice. I had proven my usefulness in the ordnance plant, so the chief asked for my deferment. But it was to no avail, for the draft board could not believe that a semieducated nineteen-year-old could be "essential" to anything—except, perhaps, a frontline infantry company. I was inducted into the Chemical Warfare Service of the U.S. Army Air Corps. My chemistry "background" had once again determined my assignment.

War as an extension of diplomacy by other means, to paraphrase Carl von Clausewitz, is at best stupid, for if real diplomacy is in process, it should be pursued to an acceptable conclusion without war. But for Hitler and Hirohito, diplomacy was never genuine; it was only a way to gain time to prepare for attack, as the blitzkrieg in France and the bombing of Pearl Harbor demonstrated. So, although I was opposed to war as an exercise in stupidity, I had no trouble joining the effort to stop the depredation being carried out by Germany and Japan. I was especially disturbed by the direct attack on noncombatants that had become generally possible with the use of airplanes in the First World War—and this concern was to haunt me after Hiroshima and Nagasaki. (Of course noncombatants have always been targets in warfare, for booty or revenge or plain viciousness, but they were not strategic targets on a routine basis until WWI.)

Even at nineteen, I was well aware that totalitarian regimes of whatever persuasion are inimical to the democratic ideal that promises (if it does not always deliver in a timely fashion) the best life we can make for ourselves. Thus I was prepared to go to war to defend that promise. I am glad I never had to face a person I was intended to kill, but if I had, I suppose I would have tried to kill him, without malice, as Caesar says in George Bernard Shaw's play, in defense of my life or of that of a colleague. But in the end, what I *did* do in the name of war, though it seemed abstract, turned out to be much more disturbing.

After the usual round of shots in Chicago, I was sent to Sheperd Field, Texas (near Wichita Falls) for basic training. It was a real airbase, but we were so far from the runways we never even heard a plane—unless one of them crashed, which seemed to happen distressingly often. I was made a drill instructor, which meant I would teach the close order drill I had learned in ROTC, with rifle target shooting and bayonet. We were "armed" with 1903 Springfields.

Once I was given a group of newly inducted Cajuns to teach. I had heard that they abhorred discipline, so I introduced myself and quoted the opening lines of "Evangeline," Longfellow's great narrative poem about the removal of the Acadians from Nova Scotia, who became "Cajuns" in Louisiana. I thought this might get me a sympathetic response from these troops, but I got no reaction at all. Evidently, they were not fans of Longfellow. One of them decided that my armband with the "DI" on it was not enough of a symbol of authority for him to have to listen to me. I explained I was teaching him bayonet because I knew more about it than he did. When he questioned this, I set up a demonstration: he was to attack me rather than the sand-filled dummies we were using. It was decidedly unfair, but I needed to make my point. I parried his attack easily and knocked the air out of him with a thrust of my own. (I had a scabbard on my bayonet.) Then I hit him with a butt stroke as he fell back. Having practiced quarterstaff play after reading about Robin Hood and Little John years before, I was a natural with the bayonet. I had never been in a real bayonet fight, but the demonstration impressed the Cajuns, and I had much less trouble with them from then on.

Advanced training, in chemical warfare, was to be at Camp Seibert, Alabama (near Gadsden). I got a book and began to learn about war gasses. They had been outlawed for "civilized warfare," but one never knew about a desperate enemy. Besides, the Japanese were said to have used them already in China, and Mussolini had dropped gas bombs on Abyssinia. I learned as much as I could with the time and resources available about the gases the U.S. had in its arsenal, the delivery methods that would be used if necessary, and the means—not very effective, I fear—that we had to protect ourselves from gas attack. World War I experience showed that gas was a chancy weapon, for if the wind changed, the attacker would become the victim. Of course, smoke and nonlethal gasses could be used as well as mustard gas, and I got my first smell of tear gas as part of training. There is a haunting painting by John Singer Sargent of gas-blinded soldiers walking with a hand on the shoulder of the man ahead that should be studied by anyone planning to use gas.

Some of our officers were poorly trained. Once I had to tell a lieutenant that he was about to fire a battery of Livens Projectors (used for throwing gas shells at the enemy) the way he had wired them, rather than testing the circuit as he intended. Since my friends were not expecting an early firing, and were scattered about in front of the projectors, we might have lost a number of trainees if he had gone ahead. I doubt these crude mortars, which looked like washtubs with thick walls, were ever used in combat by us, although the Chinese had used them (ineffectively) against the invading Japanese, for lack of anything better.

I got as near as I ever would to hostile fire in the "infiltration" course at Camp Seibert, where machine guns were fired over our heads as we squirmed under barbed wire in mud, holding the breeches of our rifles so they wouldn't foul up. Small charges were blown up around us to increase the verisimilitude. I had heard enough about this kind of combat, where troops had to advance on the ground from fixed trenches—as my father had had to do in WWI—to know that it was unlikely I'd ever see it. Still, I welcomed the chance to feel what it was like to face real fire. Of course, no one was actually trying to kill us, which makes a big difference!

For some reason, .50-caliber machine guns were assigned to us, and we learned to fire them well. One evening as I was leaving the mess, a lieutenant called to me to join him. He said some troops were rioting and needed to be suppressed. We headed across the camp, and as our direction became clear, I guessed that the "rioting" troops were from the black units in camp, which were carefully segregated. The officer wanted calm restored as quickly as possible. I was positioned behind a .50-caliber machine gun and told to fire on the troops if they came across the boundary of their part of the camp. Several other GIs were lined up with fixed bayonets when we arrived. I had no idea what the fuss was about, but I suspected that the officer's reaction was based more on prejudice than on anything that was actually happening. I could see the black troops milling around, but they seemed more confused than hostile.

Now, I am governed by rationality more often than not and could never find a rationale for prejudice, even though I was raised in Alabama and Texas in the 1920s and '30s. But there were no other "rational" people around me that evening in Alabama. The lieutenant seemed to want bloodshed and ordered me to fire, but I fired into the trees above the troops' heads. The "riot" stopped anyway, and I explained to the lieutenant when he asked that my inexperience with the .50-caliber had misdirected my aim. The next morning I was called into the captain's office to give an account of the incident, and I told him the same story. The captain grinned and said, "Corporal Hull, your firing record with the .50-caliber is excellent. However, I'm happy your aim was bad last night. The officer is being reassigned so that he will not encounter black troops."

We also trained with the 4.2-inch mortar. We could lob gas shells much more accurately with these mortars than with the Livens projectors. I found this a fascinating weapon. All the mortars I had read about in histories of the Civil War and WWI, including the German "Big Bertha," the largest caliber artillery piece in the war, threw round shot in a high, arching trajectory designed for attacking fortifications, as, after a fashion, did the Livens Projector. All these weapons were smooth bore, but the barrel of the 4.2 was rifled like an artillery piece. Its ammunition was also shaped much like a regular artillery shell, but it was dropped into the barrel from the muzzle like a mortar. Thus there had to be some different way of engaging the lands and grooves of the rifling. Ordinary artillery shells are loaded in the breech of the gun and have brass rings, slightly larger in diameter than the body of the shell, to engage the rifling when the propellant sends the shell up the barrel. What I learned was that the 4.2 shells had a skirt of brass around the back end, with the same diameter as the body of the shell and a groove facing backward. When the propellant exploded, the skirt was opened up and driven into the rifling, giving the shell the rotation that made it stable in flight. This was why it was so accurate. I remembered that the long rifles of the Revolution were also loaded from the muzzle—by ramming the patch-wrapped shot down the rifled barrel. This was a procedure about three times as slow as for a musket, but with our design, we could fire as rapidly as any other gun. There were usually four of us on a crew: one to carry the barrel, one to carry the base plate, and two to carry ammunition. My crew could set up and begin firing in a couple of minutes, or less if the officer wanted speed. Fortunately gas warfare was never introduced in WWII, so the 4.2 was converted to a small, portable artillery piece, with HE (high explosive) shells rather than gas.

Back to College

As my training neared completion, I was called again into the captain's office and told that I was going back to college, this time in the ASTP

(Army Specialized Training Program). I had pretty well committed myself to a combat role in the war by now and was looking forward to it as a new experience. I explained to the captain that I had experienced college but never combat, and that I'd like to see how I would react. But he was unmoved. I was going back to college. After a brief stop at the University of Alabama for assignment, I went to the University of Florida to be educated in electrical engineering. After a couple of terms, some of us were sent on to the University of Pittsburgh to continue the program. The rest, we understood, went to the "troops."

At Pitt, we lived in the unfinished "Cathedral of Learning," a skyscraper classroom and office building overlooking Forbes field. Although it was difficult for a GI in uniform to pay for anything in Pittsburgh, and we were always welcome at the field, it was sometimes easier to just look out our windows to see how the game went. We were on the twenty-first floor and went to class in high-speed elevators. The Air Corps trainees lived a few floors below us. As Cadet Colonel, I ordered a regular watch, and any head poking out the windows below got hit with a paper sack full of water. Once the head belonged to one of their officers, and our captain got a complaint. He confined us to barracks that weekend—not, as he said, for dropping the water bomb on the lieutenant, but for missing!

Once a friend of mine asked me to accompany him to a college graduation dance in the Webster Hall Hotel across the street. He suggested I could stay in the Boot and Saddle Bar if I didn't want to dance. I started there and met Roy Rogers (without Trigger or Dale). It must have been later that Roy got religion, because he hoisted a few while I drank a beer. I finally went to the dance and was dragged to the dance floor by one of the teachers, an attractive "older woman" of at least twenty-five. She quickly passed me off to one of the graduating students, and I met my future wife!

Pitt did not have the final term for our program, so once again triage was performed and about a third of us went on to Pennsylvania

State for the "senior" term. The program was intense, with advanced network circuit theory, radio wave propagation, radar sets (based on magnetrons and klystrons in those days), classical electromagnetic field theory, more electronics, and a series of very good laboratory courses that backed up or extended our theory work. The labs were almost impossible to complete in the time we had. Usually we could just get the circuit put together and maybe get a single measurement done before the end of the period. I figured a way around this problem. I picked the two best "hands" in the group for my partners, and as we started the lab, I worked out the theory of the assigned experiment while my partners wired the circuits. By the time they were through, I had a series of "measurements" calculated, with the only thing left to do being to set the parameters produced by the particular circuit we were using. One very careful measurement got us the parameter, and we had a fine series of "measurements" for this set up. If there was time, a second actual measurement checked my theory and arithmetic. I have no doubt the instructor knew what we were doing and probably approved of our enterprise, because we learned more this way than struggling, probably incompletely, to badly wire a circuit quickly enough to get several measurements. (I always made up for this approach by working in the lab during free time on experiments where I didn't know the answer—such as designing and building an FM receiver that actually worked, which introduced me to Bessel functions that represented the wave form of FM.) As it happened, this "get the answer" method was good training for my eventual assignment to Los Alamos. There was never time there to do things in an orderly manner. If a process worked at all, we adopted it and moved on. Improvements could be put in place after the war. A theoretical solution would be adopted for implementation without adequate experimental tests; if it was good enough, an experimental solution would be adopted whether or not there might be a better one.

On to Los Alamos

At graduation I figured I'd likely go with the troops in the Signal Corps, since I was now a communications engineer, or to the Signal Corps laboratory. In fact, I was sent to Oak Ridge, Tennessee. All I was told about this place was that it was near Knoxville. I was curious about my assignment, but no one at Oak Ridge had any idea what it was—or wouldn't tell me. I had no contact with the technical people there and was given menial jobs like mail room clerk to fill the days until I was shipped out again. It was fall of 1944 when I finally left Oak Ridge on the train. I was put in charge of a "shipment" (a half dozen men) and told to call a telephone number from a list I was given at each stop and report. All the way across the country, I dutifully called my numbers and said, "This is Corporal Hull reporting. All personnel are accounted for. No one has approached us." We got on the California Limited (well named in wartime—I think it finally got on schedule by dropping a trip) in St. Louis and ended up in Lamy, New Mexico, a tiny, dusty town in the middle of nowhere. (Lamy is still tiny, and still the Amtrak stop for Santa Fe for the few visitors who travel west by train.) I called the last of my numbers and then paced about, wondering what we'd be doing the next day—something new for sure; something interesting I hoped.

We waited in Lamy until a cowboy driving a school bus arrived to take us to our destination, which he called The Hill. Some questioning yielded the formal name of The Hill, Los Alamos. This proved to be a village of army buildings on top of a mesa reached by a narrow, winding road with steep drop offs on one side. The driver gleefully pointed out truck wrecks down in the canyons below as we made the climb. We were taken to barracks built on the edge of the mesa along what is now called Trinity Drive. Each of us was given a different place to report to the next morning. With one exception, I doubt that I ever saw any of those fellows again.

Los Alamos did not officially exist in 1943–45, the period of its most intense activity. It appeared on no map of New Mexico, its children were all born in a post office box (PO Box 1663, Santa Fe), and the bodies of deceased residents were sent back to their hometowns for burial. Robert Oppenheimer, director of the Laboratory, had chosen the site. It had been a boys' school, and some of the former teachers' and staff houses remained as "bathtub row," so named because this was the only personnel housing that contained the luxury of a bathtub. Fuller Lodge, which had been the administrative and classroom building for the school, had been much developed since the school's days and its wartime use, and it is still in use by current residents and visitors. Fences contained all the people and activities of the Laboratory. There were fences within fences, as levels of security rose as you moved inside. The outermost fence around the town and around the Lab was patrolled by mounted military police. They also manned the gates through which one could pass, either way, only with a proper document.

There was a mixture of civilian and military personnel at the Lab. Provisional engineers built buildings; military police handled external security; special engineers (like me) filled scientific and technical roles; Women's Army Corps filled the same jobs they had elsewhere in the army; and Navy officers filled administrative, scientific, and technical roles as their training and talents indicated. The civilians were scientists, spouses (of senior persons), children, laborers, and teachers (usually spouses). New Mexico citizens from as far away as Taos supplied much of the labor force and were bussed into the Lab daily. Los Alamos was more of a town than an Army base, but, perhaps, the most unusual town in the country.

I wondered if I would be put to work building a power plant or perhaps a radar system. When I reported to Major Ackerman the next morning, I noted the Ordnance symbol on his collar. He explained I was going to cast explosives! Thus it was going to be my experience at Illiopolis rather than my hard-earned electrical engineering training that was wanted here.

The major drove me out to a place called S-Site and introduced me to the lieutenant in charge of the program I was to work in. The "lab" was in a small building nearby, and there were several other GIs already in place, pouring melted explosives. S-Site was one of the most restricted parts of the installation. Just as the Lab itself was placed remotely from the rest of the population of the state to reduce security problems, S-Site was remote from where people from the Lab lived and most of them worked. This was also for safety reasons. If there should be an accident with the explosives that were all over S-Site, the principals at Townsite would not be injured! It had been explained to me that I was to lead one of the efforts here. With nine months experience in explosives, I guess I was an "expert" compared to my associates. ("In the country of the blind, a one-eyed man is king!" wrote H. G. Wells.) I had just turned twenty-one.

The castings we were making were called lenses, and it took me a bit of thought to understand what that meant. They didn't look like

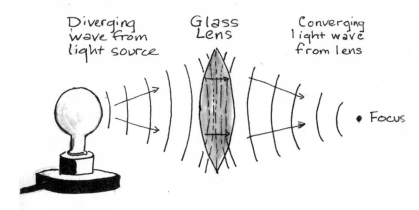

Diverging wave from light source

Glass Lens

Converging light wave from lens

• Focus

OPTICAL FOCUSING

any lenses I had seen—including those in the glasses I wore. They were binary, which means they consisted of two different explosives cast one over the other. The explosive in the inner casting was baratol, essentially a mixture of barium nitrate and TNT. The other was called Composition B, or Comp B, which was a mixture of TNT and RDX, and thus more powerful than plain TNT by itself. So one of the components was comparatively slow and the other fast in terms of burn rate. That was the clue to understanding these castings as lenses. A common glass lens focuses a divergent light field into a converging one by virtue of the fact that it is thicker in the middle than on the edges. The incident divergent spherical wave is thus slowed in the middle—because light moves effectively more slowly in glass than in air—and emerges from the lens as a converging wave. An explosion initiated at a point in a block of explosive also is a diverging wave. It is a pressure wave, more like sound than a light wave, but it still behaves the same way. The pressure wave has a characteristic velocity in a given explosive—slower in baratol, faster in Comp B. Thus a divergent explosive pressure wave started by an initiator in the middle of the Comp B surface would encounter the baratol when it had passed through the Comp B and slow down in the middle. The wave emerging from the combined casting would be convergent. Thus the combination of explosives in these castings we were making acted as a lens for explosive pressure waves.

I learned that the castings we were making would be assembled into a fifty-four–inch sphere of explosives and that thirty-two individual lenses would be fired simultaneously. A spherical explosive pressure wave would thus converge inside the shell of lenses. This clearly was an implosion. Some of the molds for our lenses were pentagonal and some hexagonal in cross section, with tapering sides. The edges of the sides were radii of the assembled sphere.

The geometry of this explosive sphere had been worked out by the mathematical physicist John von Neumann as a practical means of implementing an idea of James Tuck's, which was adapted from an idea of Seth Neddermyer's. (Tuck was a member of the sizable British contingent working at Los Alamos.) Von Neumann would go on after

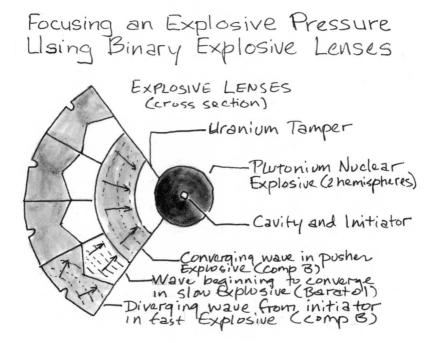

Focusing an Explosive Pressure Using Binary Explosive Lenses

EXPLOSIVE LENSES
(cross section)

Uranium Tamper

Plutonium Nuclear Explosive (2 hemispheres)

Cavity and Initiator

Converging wave in pusher Explosive (Comp B)
Wave beginning to converge in slow Explosive (Baratol)
Diverging wave from initiator in fast Explosive (Comp B)

Baratol Mold
Mold Cap

Composition B
Mold Cap

Baratol casting
after riser
machined off

Completed Casting
with fuze hole

EXPLOSIVE LENS

the war to develop one of the first large electronic computers, as well as the process of linear programming still used. (Despite his well-known postwar work on the theory of games, von Neumann was only an average poker player. I usually won a few dollars from him when I got to know him later.) It would be a while before I knew why we were setting up to make this implosion; my immediate problem was to figure out how to cast the lenses to the specifications required.

The requirements were to cast the explosives without internal cavitation or surface bubbling and with uniform mixture of the components. Most materials shrink when they cool from the molten state, thus creating cavities within the casting. To help counteract this tendency, most castings are overfilled, with a "riser" of extra material at the top that can flow down to fill the cavities as they form. When the casting is solid, the riser is machined off to leave the cast with the shape as designed. Still, it is not uncommon, even with this procedure, to find at least small cavities imbedded within the casting. Very elaborate molds have several sprues, or channels, to allow molten material to enter through several paths to reduce the cavitation and to ensure that the whole shape is filled. Our molds were not so complicated, but post-testing of the castings that had been done showed that there were still small cavities within them. In addition, bubbles on the surface frequently marred our castings . They were quite large for explosive castings—about a foot across and a foot and a half deep—so the surface may have cooled differentially, thus allowing the surface bubbles. Finally, chemical analysis of samples from the completed castings showed that the mixtures had settled out during cooling, leaving a nonuniform mixture throughout. I was told that my job was to find a way to make the castings without these three flaws.

It was an interesting set of problems. At Midland, we had simply poured molten TNT into the shells, machined in a place for the fuse, and that was it. The troubled tracer was put in separately. No one cared if there were cavities. But the lenses at S-Site were much larger, and because they were binary, they had to be cast in two stages. One explosive was cast over the other after the first was solid and its riser had been machined off. Thus they were much more complicated. But why did these lenses require so much care? I would have to figure that out for myself.

The possibility that spies would infiltrate the work at Los Alamos was a serious concern, so internal security was imposed. The guiding principle was "need to know," meaning that a worker was to have access only to as much information as he or she required to do an assigned task. Badges of different colors were worn (and still are) to indicate each person's level of clearance and the areas of the Lab he or she could visit. Those with a white all-area badge could see everything. The intent was to limit the amount of information each person had, so that if there were spies, no one of them would be able to describe the whole effort. Nevertheless, while the feared German spies never appeared, it would be revealed after the war that the Soviet KGB had been well served. There were enough spies located around the Lab that a fair description of the work was quickly available to the KGB. The most notorious of these, of course, was Klaus Fuchs in T-division (for theory), who helped me, just after the war, with understanding a technical book he had read for the author. No one, and certainly not I, suspected him. (Communism is a kind of religion and could recruit converts from any country without any problem of state loyalty. Fuchs was a German citizen of Great Britain, which only imprisoned him for some fourteen years. Compare this punishment to that of Julius and Ethel Rosenberg, who were American, and thus subject to our treason laws. No one I knew thought they should have been executed. Their treason had not affected the prosecution or outcome of the war, so life in prison seemed a reasonable penalty for their betrayal. But they suffered from the heightened feelings of wartime in the country as well as their own criminal behavior.)

These security measures were the result of a compromise. General Leslie Richard Groves, the director of the Manhattan Engineer District, of which the Los Alamos Laboratory was a part—as I would soon learn—wanted utmost secrecy, but Oppenheimer argued persuasively that an interchange of scientific ideas among the staff was necessary to accomplish the work in a reasonable time. One of the most complicated devices ever attempted was being designed and built *ab initio*—there were no precedents. The usual lag between the discovery of new principles of physics and their application to a new device was being shortened by, say, a factor of ten because the discoverers of the principles were designing

LENS MOLD

Baratol Cap

Water coils inside mold walls

Water Tubes

and helping to build the actual devices. The sharing of ideas among these scientists was essential for the rapid progress that took place. Thus, after a tussle, it was decided that Oppenheimer would control security within the Lab and Groves for the town and externally overall. (Several physicists, including Gregory Breit, who would become my PhD mentor, had persuaded the American research journals not to publish articles on nuclear physics from 1940 on, so as to give no assistance to the expected German effort. Scientific information is usually open to everyone, so the enemy knew everything we knew until this moratorium on publication was put into place. The articles were published after the war, so we had papers with a submission date of 1940 finally appearing in 1946—that's slow, even for *The Physical Review!*)

By the time I arrived at Los Alamos in October or November of 1944, the mold designs had been pretty well set and the explosives determined, although small composition changes were made as scale tests suggested. Nothing about these first weapons was carried out in the orderly way that good engineering practice demands. Parts were designed before performance requirements were determined. Complete testing of components never occurred. But the senior people (none, I think, was much over forty!) were uninhibited by routine procedures. They were creative thinkers who believed they could think their way through any problem. It is little wonder that General Groves never really understood these people. Fortunately, he trusted R. C. Tolman, an older physicist who helped moderate the interaction between Groves and the scientists. And so Groves went against a lifetime of caution, and, grumbling all the way, found all the materials the scientists needed in a timely fashion.

Piecing It Together

As an ordinary GI, I lived in a barracks with other soldiers brought in to fill the junior technical jobs all over the Lab. I didn't know much nuclear physics, but I was determined to figure out just what it was we were doing in this remote place. My security clearance at that time was blue badge and my area was S-Site. I could know everything going on at S-Site and in X-division (explosives), but not, for example, in T-division (theory) or P-division (physics-experimental). But most of my barracks mates rarely worried about these restrictions, and we regularly exchanged information about what we were doing. Thus I could always check my progress with Paul Stein, a friend from ASTP, who worked in T-division. These security breaches probably reduced the overall task time by months!

I began to put the picture together gradually from several sources. My first understanding was that we were attempting to exploit Einstein's discovery that matter is a form of energy. The famous (if not widely understood) relation is $E = mc^2$, where the energy "E" in a mass "m" is equal to that mass multiplied by the square of the velocity of light, "c." Since "c" is 186,000 miles per second, the energy content, even of small masses, is very large. I remembered how the Sunday supplements in the 1930s had had fun running the Queen Mary back and forth across the Atlantic Ocean on a cup of water to demonstrate how much energy is contained in ordinary matter. In fact, every energy source converts some mass into other forms of energy, but for most ordinary sources—your automobile engine burning gasoline, for example—only the smallest part of the mass energy of the system is converted. I figured the aim of the Manhattan Engineer District was to convert a significant fraction of mass energy into heat quickly enough to cause an explosion.

But what mass was to be converted? By this time, multiton TNT bombs had been made for the air forces of the world to drop on civilians in the cities of Europe, and Jimmie Doolittle had managed to do some damage in Tokyo with much smaller ones; so I figured we must be after a much bigger yield. I knew from my high school and college reading that the binding energy of the nucleus is a million times as great as the binding energy of atoms. We release atomic level binding energy when we explode TNT—or gasoline in the cylinders of our cars. I guessed that what we were trying to do at Los Alamos in the 1940s was release the binding energy of the nucleus.

The nucleus of the atom was discovered early in the last century, and when its structure was revealed in the early thirties, it was seen to be made up of positively charged protons and electrically neutral neutrons. Since they stay together under normal circumstances, these nuclear particles must be bound together by some force, called the strong nuclear force. It is the strongest force in nature: stronger than the electromagnetic force that holds electrons to the nucleus and hence makes possible the materials of the world; stronger than gravity, which makes rocks fall and the planets dance around the sun; and of course, much stronger than the weak nuclear force that causes radioactivity. (A fair part of my professional research career after the war would involve delineating the properties of the strong force.)

I knew that radioactivity was a nuclear process whereby the nucleus changes its structure spontaneously. Henri Becquerel had discovered that certain atomic species, like uranium, could fog photographic plates through the envelopes he stored them in. Marie Curie found other atomic species that had the same ability, and her work was my entree into the subject, as I had read about it in high school. Radioactivity is a process whereby the given nucleus emits (usually) one of three kinds of particle—alpha, beta, or gamma—and changes its character. But the energy released in ordinary radioactivity is not controllable, and the energy release rate is a property of the radioactive species, with a range from microseconds to centuries, so something else must have been our goal.

When one of my barracks mates who worked in Townsite, where most of the physics proper was done, mentioned the word fission, I had the clue to continue my search. A few judicious questions plus a little time in the library got me to a limited understanding of what we were doing. What I learned is that the binding energy per nucleon (the collective noun for either a proton or neutron) of a nucleus varies among the elements, and if we plot binding energy per nucleon against atomic number, we get a relatively smooth curve that peaks at iron 56. Iron 56 is the isotope of iron with thirty neutrons. Each element has a unique number of protons (and electrons), but an element can come in various forms, or isotopes—so called because they appear at the same place in the chemist's periodic table and have the same chemical properties—each with a different number of neutrons. Iron has 26 protons, and so has an atomic number of 26, which marks its place in the periodic table. Its mass number denotes the size of its nucleus—its protons and neutrons together—hence iron 56's is 30 + 26. The binding energy per nucleon is the energy that would be required to break up a nucleus of a given mass divided by the total number of nucleons. The binding energy of iron 56 is about 450 units, so its binding energy per nucleon is about eight units.[*]

More massive nuclei up to uranium, which has a mass number of

---

[*]Technically, binding energies are negative. If we take as zero the situation where the nucleons are far enough apart not to interact significantly, then since we have to put energy in to tear the bound nucleons apart to the zero condition, their bound condition must have been a state of negative energy. The iron case is therefore a minimum in the binding energy curve rather than a peak, but I prefer to think in terms of "minus the binding energy," and in these terms, the binding of iron peaks.

238 for the most common isotope, show a gentle drop in binding energy per nucleon that ends up one energy unit less than that of iron. Thus as we move across the periodic table beyond iron at atomic number 26, the nuclear masses of the elements become greater but their binding energies per nucleon go down. They become less stable. If such a nucleus were split in two, the resulting nuclei would be thrown into a region of greater (negative) binding energy, if the initial nucleus was at least twice as heavy as iron. The difference in binding energy between that of the initial nucleus and the sum of the binding energies of the resulting nuclei is the energy released. This is fission. The maximum difference, between, say, uranium and isotopes in the middle of the periodic table is one unit per nucleon, or about two hundred total units. The key here is that the unit is a million times the size of the units used to measure energy released in chemical reactions, so the fission of a uranium nucleus is a million times as energetic as the explosion of a molecule of the chemical TNT! And, as we shall see, unlike natural radioactivity, the tremendous release of energy in fission can be controlled.

As the nucleus gets larger with heavier elements, the total amount of repulsive force among the protons begins to make it unstable. Protons carry a positive charge and electrons are negative. Thus as the atomic number increases through the periodic table of elements, the number of protons and the number of electrons must increase by the same amount to keep the atom electrically neutral. But the number of neutrons increases, too. Crowded together in the nucleus, the protons repel each other (like charges repel, unlike attract), so the added neutrons are needed to supply enough strong force attraction to keep the nucleus together. The protons are also attracted to each other by the same strong force,* but not as effectively because it is short range,

---

*In fact, as some of my later research with Gregory Breit showed, the strong force is charge independent; i.e., neutron-neutron, proton-proton, and proton-neutron strong attractions are the same.

while the charge force, the force that makes them repel each other—called the Coulomb force after its delineator—is long range.

Physicists graph atomic stability on an isotope chart. If we call the number of protons Z and the number of neutrons N, we can plot all isotopes in a chart with Z along one axis and N along the other. Then all the isotopes of, say, carbon, will occupy a line on the chart where Z = 6 (since carbon always has 6 protons), and those of oxygen a line with Z = 8, and so on. In the region of the chart where the isotopes are stable—what we sometimes call the valley of stability—we plot a sloping line up through isotopes of elements of increasing atomic number. It is quite possible to have several stable isotopes of an element—nuclear species with the same Z but several different Ns—in this valley. For low Z (lighter elements), the valley runs approximately at 45 degrees to the axes, showing that N = Z. As masses get larger, the valley will show a bend toward larger numbers of neutrons than protons. Thus it is clear that an increasing proportion of neutrons is needed to keep the protons together.

With 92 protons and some 140 or so neutrons, uranium is the heaviest naturally occurring element and one of the most unstable. Its nuclear structure is such that it fissions relatively easily, especially U235, which has 143 neutrons. While there are other species of isotopes that fission spontaneously, as we shall see, for uranium, the nucleus has to be pushed over the brink into an unstable condition.

I learned that the Italian physicist Enrico Fermi had bombarded uranium with neutrons in an attempt to make elements heavier than uranium. Individual neutrons cannot in fact be accelerated, but nuclei can. Thus deuterium, for example, which is heavy hydrogen (hydrogen with a neutron, so that it has one proton and one neutron), can be accelerated and the proton scraped off to leave a fast neutron beam. (The electron is lost in the shuffle and has no effect on the experiment because its energy—kinetic and mass—is too small.) Early experiments with a beam of neutrons on light elements had resulted in reactions with some light particle emitted and no great energy exchange. But Fermi had found that with heavy elements, an incoming neutron was taken up by the nucleus. The nucleus then stabilized itself by emitting the neutron's binding energy as gamma radiation and became a heavier isotope of the element. This isotope could then decay radioactively by sending out an electron. Then the added neutron would become an added proton, and the target nucleus would be transformed into the element one step up on the periodic table. Fermi expected that by bombarding uranium, he would first get a heavier isotope, uranium 239, and then an element with atomic number 93—a new element, a transuranic element!

The resulting radiation in Fermi's experiment did not seem consistent with expectations, however. There were too many electrons emitted with the wrong energies. It was Otto Hahn, Lisa Meitner, and Fritz Strassman who figured out what was actually going on in Fermi's experiment. Their chemical analysis of the target material after its bombardment with neutrons revealed that the residue was comprised of elements in the middle of the periodic table. They are thus credited with the discovery of fission, though it took a while for Meitner's contribution to be recognized. The incident neutron had pushed uranium over the brink! Rather than joining the uranium nucleus to form a new element, one step heavier on the periodic table, Fermi's added neutron had broken the nucleus into two parts. It became two much lighter elements and released excess binding energy.

Further study revealed that the uranium isotope with mass 235 was the one that fissioned (or "fished," in the lingo of Los Alamos in 1943–45). If the aim of the Manhattan District effort was to use fission as a source of energy, this discovery was unfortunate in that U235 is not the most plentiful of uranium isotopes. It occurs as only one atom in about 140 in natural uranium ore. It is, by definition, not possible to separate isotopes by chemical means, and the physical means available at the time usually separated one atom at a time. It would take a very large number of separations to make a pound of U235.

I had read about electromagnetic separation of isotopes some-where along the way, and at the time I didn't know any other method. If a charged atom, called an ion, is sent through a magnetic field, its path will be bent by the magnetic force exerted on it and will curve in a plane perpendicular to the field. Because they have different mass-es, two isotopes with the same charge moving through the same mag-netic field will travel paths bent by different amounts. Thus when they come out of the magnetic field they can be collected separately—one atom at a time.

Through discussion with my friends I learned of other isotope sep-aration methods. Cream rises to the top of a bottle of whole milk (if it's not homogenized) because it is less dense than milk. If you spin whole milk about a vertical axis, the cream will collect in the center and may be drawn off there. In the very same way, a batch of fluid containing the isotopes of uranium will separate in a centrifuge, with the lighter isotope, $U235$, collecting at the center. The enriched material is then separated and recycled through the centrifuge several times, until a satisfactory level of purity for $U235$ is obtained. With only a little over a 1 percent mass difference among uranium's isotopes, this is a slow process; but it is a batch method and so is faster than magnetic separa-tion. Modern separation uses very high-speed centrifuges in the form of tubes—much discussed recently in Iraq, Iran, and North Korea!

Finally, there's diffusion. If a nonhomogeneous fluid passes through semipermeable membranes, there is again a separation by mass. A gas of UF_6 (uranium hexafluoride), made from ordinary uranium ore, will contain both isotopes. If this gas is diffused through a suitable mem-brane, the lighter isotope will pass through more easily, and thus will be enriched in the mixture collected after the membrane. The new mix-ture is passed through a membrane again and is further enriched. This can be continued until one has a very high concentration of $U235$.

This was what was being done at Oak Ridge. In fact, all three meth-ods of separating the fissile isotope from uranium ore were being used.

Large-scale plants had been built for each process. Gaseous diffusion turned out to be the most efficient method in 1944–45 and produced much of the material used in the bomb dropped on Hiroshima. It was now not surprising to me that the people I came in contact with during my ten days or so in Oak Ridge were unable to tell me anything. These were uncommon activities, outside their experience. I had seen the plants but had met no one who could tell me what went on inside them. There were no technical staffers where I was located during my brief stay.

Fermi's scheme for making elements beyond uranium was not wrong; his result was just obscured initially by the fission of U235. The heavier isotope, U238, does indeed absorb some of the incoming neutrons, becoming, for a moment, U239. This isotope then decays by emitting an electron and turning one neutron into a proton—but it does so twice in succession! The result is an isotope with $Z = 94$, a new element *two* steps above uranium on the periodic table. This new element was named plutonium, and the intermediate element neptunium, since the planets Neptune and Pluto lie beyond Uranus. The chemical symbol for plutonium is Pu, and the isotope Fermi made was Pu239.

I had not met Fermi at this time, but he was already legendary to us at Los Alamos for his ability to see through problems and offer solutions. It was said that one rarely wished to oppose Fermi's ideas, not because he would object, but because you'd probably be wrong! I met him after the war when I worked in T-division, and I found him modest and very open to young physicists (if I could call myself a physicist then). He once said that the future of physics was with youngsters, as the older physicists (he was in his middle forties then) had had their time. I should have been happy to have a fraction of the ideas he had left in 1946.

It turned out that Fermi's transuranic Pu239 fissions more easily than U235, so it is an explosive as good as or better than U235. Fissile plutonium has an important added advantage over fissile uranium in that it can be chemically separated from uranium. Their chemistries, though similar, are different enough. To make a piece of plutonium, then, one has only to bombard enough uranium with enough neutrons and separate out the plutonium. The question was how to bombard the piece of uranium with the neutrons.

By 1942, it was known to the international community of physicists that the fissioning of uranium not only yields energy, but also two or three neutrons. This means that if there are enough uranium pieces in close enough proximity to each other, one fission could produce another, and another, and another, as the cascade of neutrons from one fissioning atom would bombard the next. This is called a chain reaction. Nature, as always, puts in a complication: the neutrons from fission are fast, and for a controlled reaction the fission of the suitable isotopes of uranium and plutonium must occur only with slow neutrons. Thus if a factory to produce plutonium was to be made, the fission neutrons in a chain reaction would have to be slowed down—by a lot. The factor needed is about 25,000,000!

It has been known for two hundred years or so that a collision between particles results in a loss of energy from the incident fast particle. If the target particle is stationary and of mass equal to that of the incident particle, the loss is maximum. Thus if a fast neutron were to collide with some moderator (as the element chosen to slow the neutrons is called), it would lose energy and hence velocity. Ideally, the mass of the fast particle and the stationary moderator should be equal, for then it would be possible to halve the energy in each collision. But the lightest readily available solid material that could be made pure enough was carbon, in the form of graphite. (A solid would contain enough moderating atoms per unit volume to cause enough collisions per second.) Carbon's mass number is fairly low at 12—not as efficient as mass 1 would be, but, perhaps, acceptable. Purity was essential. The cross sections for neutron absorption were not then known for every substance, and the design of a machine would depend on knowledge of the absorption rate for whatever moderator was used. As few neutrons lost as possible was the goal! The fissile isotope in the uranium ore was the scarce one, as we have seen, and if too many of its nuclei were missed by the slow neutrons, the chain reaction would not be sustained. It has been said that an error in the chemistry of graphite by a German chemist, suggesting too much loss of neutrons because of impurities, was one of the reasons there was no successful German nuclear project during the war. An overestimate in the amount of nuclear explosive needed for a bomb, attributed to Werner Heisenberg, was also said to be a cause of the failure of the German project.

Fermi and another refugee from fascism, the Hungarian Leo Szilard, proposed a machine to put together all the conceptual pieces for plutonium production, in which uranium fuel would be mixed with graphite as a moderator. Once started, the mixture should continue reacting on its own. The heavy isotope of uranium would absorb a neutron and eventually turn into plutonium; and the light uranium would also absorb a neutron and fission, yielding two or three fast neutrons. The graphite would slow down the fast neutrons so more fissions would easily follow. In this operation, the fissions were sources of neutrons to make plutonium from U238, not for energy production, although some energy is always produced in a chain reaction system. After a period of operation, the machine would be turned off, and the depleted fuel would be treated chemically to separate the plutonium. The element cadmium absorbs neutrons so readily that a small amount inserted into the machine would stop the chain reaction and turn off the machine.

Sounds straightforward, doesn't it? But, of course, no one had ever tried the process. Before coming to Los Alamos, Fermi built such a machine as a proof-of-principle experiment in the squash courts of the University of Chicago (under the unused football stadium seats!). A layer of graphite was followed by a layer of uranium slugs interspersed among graphite pieces, then another layer of graphite, and so on. Wooden sticks clad with cadmium were inserted periodically for control—they could be pulled out of the pile to allow the chain reaction to go. The layers were piled up in a roughly spherical mass to make calculations easier—hence the name "pile" for the first machine, now called a reactor. Since they are chargeless, neutrons can travel quite a distance before being absorbed, so there was a critical size. If the pile were too small, too many neutrons would escape and the chain reaction would die. Of course, a number of properties of uranium, plutonium, and graphite, plus those of any structural materials around, had to be figured out before the critical size could be calculated.

In November of 1942 the pile grew each day as assistants built up the layers. They would complain that no amount of washing got all the carbon off their bodies, but they stuck to the effort. As the number of layers approached the design level, the cadmium-clad rods would be pulled out and the neutron production measured. When the production was such that one fission occurred for each previous fission, the pile would be in a self-sustaining mode. On December 2, 1942, the test in the morning was promising. Fermi likely knew that the next layer would take the reactor critical (i.e., bring the structure to the size and configuration that each fission caused one new fission)—so he called lunch! The reactor went critical that afternoon. Fermi and his dedicated crew had produced the first ever sustained, controlled fission chain reaction. A triumphant coded message was sent to waiting scientists in Washington: "The Italian navigator has landed and the natives are friendly." Fermi once said to me that he thought the message a bit theatrical and a bit too focused on himself, but it was an unbreakable code.

The building of a successful nuclear reactor the first time it was tried is a testimonial to the genius of Fermi—who, as theoretician and experimentalist, was of a vanishing breed—as well as of Szilard. It was typical of the wartime effort that there was not enough time to try out parts of the machine; the best one could do was to make sure the most important elements were understood. The engineers involved in planning the pile experiment were uniformly cautious, claiming the job could not be done in the way Fermi and Szilard proposed—and probably couldn't be done at all! But the physicists were undaunted. They just needed a few important parameters, a few good ideas, a crew of dedicated workers (usually graduate students in physics)—and a lot of luck. The pressure to make this experiment work was enormous, for if such a reactor could not be built immediately, then nuclear bombs would probably not provide a decisive advantage to the Allies. There would then be no plutonium explosive, and we would have to depend exclusively on the much less efficient process of isotope separation at Oak Ridge for U235. In Europe, Britain was recovering from the sustained German air attacks of 1940–41, and German General Erwin Rommel, Hitler's leading field commander, was still in Africa, though the Afrika Corps had been stopped twice at Al-Alamein. In the Pacific, the Japanese had been defeated in the Coral Sea and at Midway, but they had beaten the British at Singapore. Their war capability was far from erased, and they continued to expand their control over the region. The outcome of the war was clearly still in doubt. We needed a decisive weapon.

The success of the Chicago pile meant that the major theoretical physicist, Eugene Wigner (whom Breit had helped get located after Wigner escaped from Hungary), could now go ahead with the building of reactors for the production of plutonium near Hanford, Washington. His task was to produce several critical masses of plutonium at about fifteen pounds each. The Hanford site was chosen so that the cold waters of the Columbia River could be used to cool the reactors. Even a reactor as crude as Fermi's pile makes heat, and it must be dissipated if it is not to be used. If a nuclear reactor produces too much heat too quickly it will undergo a meltdown, as we have learned since the war. I thought the Hanford project should have been named for the Augean Stables and Wigner given the code name Hercules. That was a little too mythical for the army, but Wigner's task was indeed herculean: to turn Fermi's proof of principle into a production facility, with each reactor much larger than Fermi's. There was no time for an orderly transition from development to production, for design of the plutonium bomb was already underway—and had been since 1941.

So now I knew how we expected to get the nuclear explosives. Oak Ridge was to separate the U235 and Hanford produce the plutonium. It was left to Los Alamos to design and assemble the devices. Both uranium and plutonium were needed, I understood, because it was not really known whether either explosive would work. We had better have two chances to succeed. Uranium would be easier to work with, but plutonium would be easier to make—assuming it could be made at all—and we would need much less of it. Plutonium fissions more easily with fast neutrons than does uranium, so the critical mass needed for plutonium was nearly three times less than that required for uranium. Thus, if a number of bombs were needed, they would probably be plutonium-cored.

In order for a reactor to become self-sustaining, it has to have a critical amount of fuel in an optimum physical arrangement, as Fermi and Szilard demonstrated. Control has to be exercised to keep it at the critical multiplication rate, which is exactly one new fission for every prior one. But a nuclear explosion would require that the system become *supercritical*. In other words, there would have to be a very rapid increase in the multiplication rate after the initiation of the chain reaction. At the same time, we would have to make absolutely sure the reaction didn't start before we were ready.

Now I could conceptualize what it would take to make a nuclear bomb. First, nuclear explosive as near pure as possible was needed. Second, we would have to know the critical mass (and so a new cliché was introduced to the languages of the world!) of the explosive to be used. Critical mass for a sample of explosive is essentially the same idea as critical size for a reactor: it must be great enough to retain the neutrons within the reacting volume until they have caused a new fission. In addition, for a violent explosion, one needs the density of the nuclear material to be high enough that an exponential increase in the multiplication rate takes place. This is supercriticality. Third, we would have to figure out how the critical masses were to be assembled and the explosion initiated when the bomb was dropped—and not before!

I understood that one way to achieve the critical mass and density of the nuclear explosive was to force two subcritical masses to collide. If the collision is violent enough and the critical mass has been calculated correctly (obviously it depends on the fission rate of the nuclear explosive for neutrons in the environment of the colliding pieces), then supercriticality should occur. One way to produce the desired rapid assembly is to shoot, with ordinary chemical explosive propellants, one subcritical piece as a projectile into another as a target—as in a gun. The design of the first nuclear bomb ever used in war was just such a gun bomb, and the explosive was purified U235. Once the Navy gun designers were convinced that the gun did not need to be built to work more than once—and so did not require the usual massive structure—a bomb that a B-29 could carry was designed. There was no test of the uranium bomb before its use. U235 was too scarce, and the designers had confidence in its simplicity. This bomb was named Little Boy because it was diminutive compared to its plutonium companion. (Weird stories that the name was chosen to diminish its killing power in the public's eyes are nonsense.) It would be dropped on the Japanese city of Hiroshima.

Plutonium presented an unexpected problem. Not only Pu239 is made in a reactor—so is Pu240. It turns out that the latter fissions all by itself! Unless the subcritical pieces could be kept far enough apart, it would be impossible to use a gun assembly of plutonium from a reactor, because the neutrons from the spontaneously fissioning Pu240 would cause preignition. It turned out that there was no propellant fast enough to assemble the pieces before preignition if they were far enough apart to prevent it ahead of time. Nuclear times are measured in millionths of a second, and propellants operate at thousandths of a second. The plutonium gun that was designed was long and thin and, in honor of a popular film series of the day, was called Thin Man. It was never built.

The solution to the plutonium problem came from two members of the design team: Seth Neddermyer, and, from the British contingent, James Tuck (with some design help from von Neumann, as we have noted). Neddermyer thought that ordinary high explosives could be used to crush a nuclear charge to supercriticality if a number of pieces were fired around the nuclear material simultaneously. He was able to prove the validity of his idea with a cylindrical geometry, but not with the precision needed. Tuck's idea was to cause an initial implosion by arranging explosives in a spherical geometry and somehow focus the shock wave on a hollow (and thus subcritical) sphere of plutonium. Focusing shock waves would require lenses—hence, explosive lenses! Now I knew what my lenses were for and why I was asked to cast them with such small tolerances. This design of the plutonium bomb, given its profile, was aptly named Fat Man.

Casting Lenses

It had taken me only a month or so to pick up the information I wanted to put our work at S-Site in its scientific or engineering context, and, of course, our casting work went ahead as fast as trial and error would allow. There was little time. The D-day landings in France had been successful, and Allied armies were advancing with few setbacks, so the end of the European war was in sight. If we had had an operable bomb of either design in the winter of 1944, it might still have been used in Europe to shorten the war and thus reduce casualties, but we weren't ready. The Pacific war was also going our way by then, but we could not yet be confident of finishing it without an invasion of Japan—with unthinkable casualties. We needed an alternative.

Our mandate was still to get the "gadget" working as quickly as possible. All available machinists were making molds in the shops at Townsite as fast as they could. George Kistiakowsky, the group leader of X-Division (for explosives), and hence of S-Site, would never feel that we had enough. Our

job was to use these molds to make the "perfect" castings that were needed for a successful implosion. They would be arranged in a sphere, detonated simultaneously, and by their structure they would converge the
explosive wave inward to a point. Inside would be the subcritical mass
of Pu239 in the form of a hollow sphere about the size of an orange and
made up of two hemispheres. It is a mark of the confidence of the physicists involved that there was never time to test an actual critical mass of
Pu239, although experiments could be done to improve the calculations. This most important design parameter was extrapolated to the size
of the hemispheres the metallurgy division was to fabricate.

The hollow at the center of the plutonium sphere was to contain the
initiator, a strong neutron source. Called the urchin, in reference to the
sea urchin's shape, it was a roughly spherical body with an interior cavity lined with teeth. Foils of polonium and beryllium would be clad to
opposite sides of projections into the hollow center of the plutonium
sphere. This arrangement would prevent the polonium and beryllium
foils from touching each other, therefore interacting to make neutrons,
until the implosive wave crushed them together. The plutonium hemispheres would be surrounded by a thick, natural uranium tamper to
contain the neutrons. (It was the densest metal available, and the added
fissions were welcome.) Outside the uranium tamper would be a layer
of plain Comp B, arranged in a sphere concentric with the inner
spheres. This was the "pusher" layer of explosive, intended to speed up
the explosive wave. Outside the pusher would be the sphere of explosive
lenses my crew and I were scrambling to perfect. The nuclear explosion
would occur when the explosive wave crushed the subcritical plutonium hemispheres together into a supercritical mass—at perhaps twice
the density of ordinary metallic plutonium. This wave would originate
at the fuse points at the top center of each lens and converge precisely
onto the two hemispheres—and converge uniformly at that!

While no one ever explicitly pushed me—I suppose it was clear
we were working as fast as possible—I understood by this time that the

project waited on lenses of the quality described to me by Major Ackerman, and I felt the pressure. Ordinarily, I get along with people who work with me, but once when one of the powder men thought he needed a rest from our work, I crushed him verbally. I think I was tired from lack of sleep, but since the berating was unusual, it was probably more effective. After that, no one else seemed to need time off!

Later I was to learn that our lens molds gave Kistiakowsky ("Kisty" to us) the largest problems he faced. His first difficulty was to figure out what explosives to use. The "science" of explosives, as we've seen, hardly existed at the time, but Kisty knew much of what was known. A Harvard physical chemist, he had invented Comp B. The burn rate of the explosives available could be used to set the ratio of slow to fast explosive components, and that ratio in turn would determine the lens design, just as the ratio of the speed of light in air and in glass determines the design of optical lenses. Mixing explosives allowed some control of the burn rates. Kisty knew he'd have to use existing explosives, so with as much testing as time allowed—mostly just to fix burn rates—he had settled on Comp B and baratol just about the time I arrived in Los Alamos.

We SEDs (members of the Special Engineer Detachment) were working diligently to solve the three problems I had been set: eliminate the bubbles on the lens surfaces, eliminate the cavitation within the lens, and keep the components of the two explosive mixtures from separating as the mixture cooled. The surface bubbles were the easiest to fix. The molds for the lenses had a double wall containing water coils along the inner surface. Our explosives had such low melting points, we simply melted them in candy kettles. It was fairly easy, then, to pump hot enough water through the coils to keep the material flowing as it coated the walls of the molds. We could lower the water temperature gradually as the casting cooled, and no bubbles should form. It took us many trials and errors to find just the right rate of cooling. It was finicky work, requiring us to sit over the molds and continually adjust temperatures as the explosive solidified, but once settled, the correct schedule

was maintained throughout the rest of the work I was associated with. I won't say all our castings had perfect surfaces, but they were very smooth! Kisty no longer needed a dentist's drill to fix the surface bubbles. Earlier, when the bubbles had seemed unremovable, he would drill each one out like a dental cavity and fill the hole with explosive.

The remaining problems of cavitation and mixture separation took a while longer to solve. I tried stirring the slurry through the riser. This seemed to help, but it still wasn't satisfactory. Finally, it occurred to me that we might stir the slurry mechanically while it was setting. I had been a waiter in the College Grill at Mississippi State to earn my board at twenty-five hours a week for twenty-one meals. Among the services we offered our customers (fellow students) was making milk shakes for them on order. I specialized in smooth, well-blended shakes, with the ice cream fully permeating the fluid. To accomplish this, I used a Hamilton-Beach stirrer. I reasoned that I should be able to get the same result with the explosive slurry if I had a proper stirrer. We were already machining off the risers of our castings with air driven drills (spark-spew-ing electrical machinery is not allowed on an explosives casting floor!), so I thought I might get longer bits with propeller tips and try my scheme. The idea was accepted so readily I thought it might have been discussed before. We got our stirrers very quickly.

The application of my idea was not quite trivial—which is typical of most of my schemes, I am told. The problem was that the drill bit had to be withdrawn ahead of the freezing inner surface of the solidifying explosives. To find the shape of this surface and anticipate its rate of contraction, I thought we could pour as many molds as we had and then pull the partially solidified castings out at regular intervals, such as ten, fifteen, twenty minutes, and so on. The remaining molten slur-ry could be poured out of the riser. After sawing the partially set castings in half vertically, we would be able to see the shape and dimensions of the cavity that remained inside the partially frozen casting. Sawing explosives may sound dangerous, but with a monel metal (soft copper alloy) saw, there was no trouble—although I did all the sawing myself just to make sure. We were already using monel bits with which to machine off risers without any problem.

Once the cooling profiles were determined, we could then set the stirrers to be withdrawn just ahead of the solidifying explosive. It occurred to me that this method should solve both of our remaining problems simultaneously. The mixture would be kept uniform while the cavitation was suppressed by the stirring. Our experimental castings were cut open and samples cut out for chemical analysis to see if we had solved the mixture problem, and they were visually inspected for cavi-ties. A more thorough test of the cavitation question was done with gamma ray photographs. Ordinary X-rays wouldn't penetrate the barium in the baratol; besides, gammas were plentiful, as Los Alamos had on hand what must have been most of the world's supply of radium.

Things were going well with the experiments, so I was able to get some leave for Christmas 1944. I was anxious to know the results of the

new stirring protocol we had just developed, but the chemical analyses would not be available before I left. So I arranged for my deputy to call me when the tests were finished. Since the results would be numbers I could interpret, he was to tell me about the birth of his sister's baby—its length, weight, and time of arrival. I'm not sure that such a baby would have looked human, but if any Army security censors had been listening in, they would not have guessed we were discussing analysis of castings. The results were good, so the pleasure of my visit with my family and my fiancée was considerably enhanced.

When I returned, we had a casting protocol that produced usable lenses. Here is what we did: First we prepared the mold by heating it internally with the water coils. Then the casting cap was inserted. This was a metal cover for the mold that determined the size and surface shape of the casting. It had a cylindrical riser in the middle, through which the molten explosive was poured. The extra explosive in the riser insured a supply of material to keep the level correct when the stirrers reduced the initial volume by collapsing the cavities.

When the mold was ready, or as it was being prepared (we had several hands), the baratol would be melted in one of our candy kettles and poured through the riser hole from carrier ladles.

As soon as the baratol was poured in, the protocol began: Change the mold wall water temperatures according to our experiments; start the stirrer and raise it according to the timing we had determined; when the casting was solid, remove the cap and replace the stirrer propellers on the drill press with cutter blades; and then cut off the riser in the shape that had been calculated for the final casting.

Machining the casting in the mold insured that it would retain its dimensions longer as it continued to cool.

We repeated the process for the Comp B. A new cap in the shape of the outside sphere, with a riser, was put in the mold, and the Comp B was poured directly over the baratol casting that had been machined to the proper shape.

When the Comp B protocol had been finished and the casting was solid, the cap was removed, the stirrers replaced with the final cutters, and the lens top machined to its final shape, a section of a sphere.

We received few visitors at S-Site, and I never knew whether anything we did was known to others. Only recently did I learn, in a piece written by a then Navy officer assigned to V-Site, that our methods were used by others who were also making castings. V-Site was where our lenses were taken to be assembled into their final form. It was adjacent to S-Site, and, in fact, is incorporated into S-Site today. I never had time to visit it. After the war, I met a metallurgist at Yale who told me he had been trying to prevent cavitation with a method called pelleting, where solid pellets of Comp B are mixed into the slurry to form freezing centers. He had not been successful, but I did not know of that effort—or any other—because there was never time to talk to others around the site. Some other group cast the one-half scale lenses that were used to test the lens idea (without, obviously, any nuclear explosive, but with radio lanthanum to check the symmetry of the implosion). After we cast a shot's worth to demonstrate our technique for them, we never heard any more about their work. (Being small, they were easier to cast anyway.) The same thing happened with the solid pieces of Comp B that were used as the pusher explosive layer in the final assembly: we cast a few to see how it could be done without settling or cavitation, and others cast for production.

We did receive a visit once from Oppenheimer and General Groves, just as we were learning to cast good lenses. As was typical of Oppie, he did not ask me to describe what we were doing but launched into an accurate description himself. He always knew what everybody was doing. Greasy, as we alliteratively called the General (he was overweight and carried candy bars for sustenance between meals), was all attention. As he moved closer to the mold to see what Oppie was talking about, he stepped on one of the water lines we used to moderate the cooling. Naturally the line popped off at the wall connection and water at about the boiling point spewed across the lab, striking Groves in the biggest target—his derriere! Since I valued my stripes I managed to hold my laughter as I turned off the faucet. But when Oppie turned to me and said, "It just goes to show the incompressibity of water," I broke up. Luckily Groves was too embarrassed to notice, so I kept my stripes.

We finally attained a reasonable system for casting lenses sometime at the end of 1944 or early 1945, but if I expected a relaxation of pressure after that, I was wrong. Lieutenant Hopper called me into his office after a few days and said, "We are bringing in a group of powder men from New Jersey for production runs. I want you to be foreman. Of course, you'll need to continue supervising development in case better methods can be found, but I'm satisfied that we are ready to go into production. There isn't time to train someone else to be foreman."

Germany was finished by this time, having been beaten decisively in the Battle of the Bulge. (They had effectively been defeated since the breakout from Normandy, but it took five months from Bastogne for the die-hards to realize it.) Allied forces would be stopped by Eisenhower at the Elbe River, short of Berlin, to save soldiers' lives. Berlin was doomed anyway by the Russian troops invading Germany from the east. Saving Allied lives returned as a policy in the decision Truman was to make a few months later. But the war in the Pacific was still costing Allied lives every day. Japan was taking a heavy toll as well, with the new fire bombing, but in 1945, that was not a consideration in America. Of course, none of this affected my response to the "request" that I be foreman. A request from an officer is an order to a sergeant, as I now was. Besides, I thought it would be interesting, which was always an inducement to me.

I knew the task before me. It took
thirty-two lenses to make a full set,
twelve pentagonal in cross section,
and twenty hexagonal. Our job would
be to make at least three sets that met
Kisty's criteria for acceptable lenses.
One would be used for a blank test,
the first test of a complete set of full-
scale explosive lenses. Then the whole
gadget would be tested down south at a
site within the White Sands Missile Range near
Alamogordo, New Mexico. (The actual site is nearer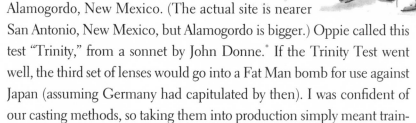
San Antonio, New Mexico, but Alamogordo is bigger.) Oppie called this
test "Trinity," from a sonnet by John Donne.* If the Trinity Test went
well, the third set of lenses would go into a Fat Man bomb for use against
Japan (assuming Germany had capitulated by then). I was confident of
our casting methods, so taking them into production simply meant training the new powder men in our protocols and equipment. We would
have all the molds available and a dozen men to work with them.

When the powder men arrived at the casting building at S-Site, I
greeted them with a quick discussion of our protocol for the lenses. Soon
the leader of the group—all men about twice my age—stepped forward
and offered me a chew of tobacco. I was aware that powder men
chewed—it's the only safe way to imbibe tobacco in the presence of
explosives (and the permanently wet floor made spittoons unneces-
sary)—so I was prepared. I said, "What do you chew?" and the reply came

---

*The fourteenth "Holy Sonnet" in H. J. C. Grierson's canon. The first line is
"Batter my heart, three person'd God—" and the tenth line is "But I am betroth'd unto
your enemie," which may suggest why Oppie chose the name. Gregg Herken gave me
this reference when we shared leadership of a Smithsonian tour.

back, "Mail Pouch." I said, "Thanks anyway, but my chew is Brown Mule. Have a chew of mine." I knew I had them by the looks among the men: if this kid chewed Brown Mule, he had to be a real powder man! That, of course, was what had I hoped for. I had never chewed tobacco in my life until I practiced a few days before the men arrived. I chose Brown Mule because it was the strongest chewing tobacco available in the PX. I had to chew it for about six months, but it was worth it. These more experienced men worked very hard with me and never hesitated to take direction when I needed to give it. I never knew exactly where they came from, but I remember hearing they had worked at the Dupont works in New Jersey. There was little time for personal exchanges, but I did find out that they had about ten times my two years experience with explosives, though not the ones we were using. I certainly didn't need to teach them about safety. You don't last twenty years in the explosive business without learning how to handle the stuff carefully. This may well have been the reason that civilians were brought in for production rather than inexperienced GIs. Their knowledge of the basic properties of explosives and the teamwork necessary for production saved us a lot of time—and, as usual, we didn't have any!

We usually had no problem getting any supplies we needed. But once we were low on Comp B, and while there was a shipment waiting in Albuquerque, the roads were so bad the regular truck drivers were not asked to pick it up. At 6,000 to 7,000 feet elevation, snow in northern New Mexico is no joke, and explosives are not a welcome load when the roads are bad. (The roads at S-Site itself were pretty rough, and there was even some concern about driving loads of high explosives over them at all. I heard that once when General Groves visited Los Alamos, Kistiakowsky took him to S-Site in a jeep in which the springs had been made inoperative with wooden blocks. As a result of that trip, those roads were improved!) I got a truck from the motor pool and went to Albuquerque. The trip there was uneventful, if a little tense. Coming back, I was concerned about La Bajada, a famous hill half way to Santa Fe from Albuquerque. The road was a dangerous switchback when Route 66 was opened in 1926—a real adventure for travelers. By 1945, it was still steep but relatively straight (today it has been tamed to Interstate standards as part of I-25) and well enough traveled that there was no snow or ice to contend with. So all went well—until I had just gotten through Santa Fe. The road to Los Alamos climbs out of Santa Fe. Back then, it was a two-lane black top, and I remember it as steeper than it is now and in a slightly different location. My truck was well identified with red flags sticking out from each fender, and the Santa Feans knew to keep away. Over the top of the rise, coming toward me at a good clip, appeared an old Ford flatbed truck piled high with crates of live chickens. They were tied to cleats with ropes that looked a bit thin to me in the instant I had to see them. As the truck came over, it began to fish tail. I assume it had hit a patch of ice. By very rapid calculation, I figured that its third swerve should put it into my left front wheel, so I took my truck just to the centerline and hit my horn. My intention was to scare the other driver away, somehow, by calling attention to the fact that I was carrying explosives (several tons, as it turned out). My ploy worked; the other driver pulled his truck to his right and we passed easily. In the rearview mirror I saw the truck go into the ditch with much dumping of chicken crates, but I'm sure the driver preferred that outcome to involvement in an explosion of Comp B. I never pass that hill out of Santa Fe without remembering those chickens.

We worked very hard that spring. We had so few molds that three shifts were necessary, and I began sleeping on my desk or the floor so as to be available for consultation. S-Site mess hall served late dinner at 11:00 p.m. and early breakfast at 3:00 a.m., and I frequently ate five meals a day—without gaining an ounce! (The menu is still served in a restaurant in Tesuque started by the wartime boss of our mess hall.) I slipped into Townsite for a shower and change of uniform as appropriate, but most of my days were spent at S-Site.

A warning system alerted us to approaching thunderstorms, which are a daily occurrence in New Mexico in spring and summer. The safety rule was to shut down casting when the storm was coming our way and wait it out in a nearby bombproof shelter. I thought the heavy copper spikes sticking up every few feet along the crest of our building and

anchored to pits of copper sulfate for good conduction were safe enough; so when the call came, I simply closed the shades of the casting building, and we continued work. Whenever a storm hit us directly, the lightning display was enough to keep my colleagues inside anyway!

One day after the war had ended, the chief safety officer invited me to lunch. I was quite surprised, since we hadn't been friends. If anything, we were polite opponents. My theory of safety was that if I knew the risks thoroughly, and what to do to reduce them, I could do whatever was needed to advance the project without considering rules. I took the fact that there were no accidents during my watch at the casting building as support of my theory, but professional safety people did not take kindly to it. We had a pleasant lunch, talking over wartime work at S-Site, and just as we were breaking up, I discovered why I had been invited. The safetyman was leaving to go back to his prewar job. He looked me in the eye and said, "I am sure I shall never again be so pleased to leave anyone as I now do you!"

There was one accident at S-Site, and it had nothing to do with explosives. I had taken my assigned staff car to Townsite to pick up a sensitive instrument I had ordered for the lab. Driving back, I took a left turn too smartly, and the instrument began to slide to the floor. Holding the wheel with one hand, I reached over to secure the instrument, and looked up to see one of the steel towers that carried steam lines over the road to the casting building coming toward me. I hit it squarely, still holding the instrument with my right hand, and broke the steering wheel off its post with my nose. Someone came out, took the instrument to the development lab at my insistence, and then took me to the post hospital. The surgeon said, "That's not so bad," inserted a steel rod up my left nostril, and pulled! "There," he said, "it's straight now." Perhaps it was, but I still have difficulty breathing through the left side of my nose. I heard that the captain of our SED company wanted to break me to private and prevent me from driving, but the major must have intervened because I kept my stripes—and even got a rocker (a curved stripe

below regular chevrons of sergeant) signifying the equivalent of a staff sergeant. I was assigned another staff car next day.

The Trinity Test

Our perseverance paid off, and by July of 1945 we had an adequate stockpile of acceptable lenses. I was looking forward to seeing how the Trinity Test would go. I had nothing to do at the test itself, so I assumed I would not be able to see it. I certainly wasn't a VIP, but Kisty was, and he must have told the right people I was to get a ride to the Trinity site and join the real VIPs—Edward Teller, for example, and Groves. I was hoping the General (two stars now) would not remember me from the dousing he got at my lab, and so it seemed.

In the tense days leading up to the test there were three questions that dominated conversation. First, we wondered if the bomb would work—that is, if it would make a nuclear explosion. The implosion bomb is a complex device, and its correct operation depended on all components working as designed—not as tested, as there were few prior tests. If each component of a device has a one in a million chance of failing, and you have a million essential parts, statistically, you are guaranteed a failure. The odds weren't that bad in our case, but complicated machines depend on all parts working. (The later failure of the space shuttle Challenger due to a sealer that got stiff when cooled is an example of how easily the most carefully designed devices can fail.) We were probably not as sophisticated as modern process engineers, but we understood something of this kind of analysis.

There were several things that could go wrong: The lenses could fail to put a symmetric pressure wave onto the plutonium nuclear explosive, because, for example, the firing mechanism—which was at the limit of timing technology in 1945—might not set each lens off within a microsecond of each other. Or the lenses might not fit perfectly. We had had to compensate for shrinkage by putting shims between the castings. This was done with the pusher blocks as well. What if the shims put spikes in the pressure wave as it hit the nuclear explosive? Suppose the pressure wave was symmetric and sufficiently powerful to crush the plutonium, but the initiator didn't work as designed. It was possible, despite the careful calculation of the expected hydrodynamics, that the plutonium would blow apart and stop the chain reaction before completion, giving a smaller explosive power than expected. It had been planned that the valuable nuclear explosive would be recaptured if there were a chemical explosion but not a nuclear one. A large tank, called Jumbo, had been designed to be placed around the gadget to catch the unexploded nuclear core material in such an event. But Oppie was confident enough not to use Jumbo. It stands today outside the fence for the test site with holes blown out of the ends—the result of a postwar "experiment" the engineers did to justify its high fabrication cost.

We were teetering between confidence and uncertainty. The most able people in the world were working on the gadget, but much of their theoretical work could not be tested. The best experimental physicists had worked on the details and done what experiments time allowed, but they knew that many details were not pinned down. The younger people like me listened to the leaders and drew our own conclusions. And so it was that among the people who had some insight, bets were placed on what the yield would be, and they ranged from dud to 20,000 tons of TNT equivalent. (This was the energy unit we used, as it was more intuitive than millions of joules). I bet $10 on at least 18,000 tons yield. This was below the design yield, but I was thinking that the core would blow apart before all the nuclear explosive had been consumed. I was to learn later that the experts, too, didn't expect all the nuclear core to be consumed, and so the size of the core mass reflected this. I was close enough, in the end, and won my bet.

The second question—after whether the gadget would work—was whether a successful blast would set off nuclear reactions in the atmosphere and burn the Earth to a cinder. It was a possibility, and I learned that Fermi was offering odds in favor. Typical of his droll humor, he never said how he would collect if he won! However, Hans Bethe (and I understand there were others) had done the calculations and said there was no chance. Now Bethe ("der Hansel" to us youngsters, a nickname he was delighted with when I told him about it forty years later) was one of two or three major theoretical physicists in the world. He won the Nobel Prize ten years after the war, but the work that merited it had been done by this time and was already recognized by his colleagues. He had shown how the sun made its energy from a series of reactions resulting in the fusion of hydrogen into helium (involving binding energies again, but at the beginning of the periodic table rather than the end). I'm sure everyone—even Fermi—was willing to trust this senior theoretician's reassurance when it came right down to it.

Of course, it was the lenses themselves that were of greatest concern to me. I knew that the blank test was supposed to have preceded the Trinity Test. This was the test specifically of the lenses, where a full-scale assembly of one of the sets of thirty-two lenses we had cast would be exploded. Instead of the nuclear core, instruments would be placed inside the sphere to measure the force of the implosion. I hadn't yet heard the outcome of the test, or even if it had taken place. When I arrived at the Trinity Test site I learned that it had failed! I wondered why we were still going ahead with a test of the whole gadget. I had been as confident as one could be that our lenses were good. There had been nondestructive tests of each one individually, and some of ours had been rejected, but this is what one expects on any production line, and we had kept the rejections minimal. But suppose we had missed some fatal flaw. I couldn't think of what that could be other than a major design failure, and my confidence was shaken. As I learned a bit later, the Trinity Test was going ahead because Bethe had looked at the blank test data and shown that it was the instrumentation itself that had failed. Although the lenses might still have been defective, or improperly designed, the failed test was no proof that they were the problem. Kisty had convinced Oppie that the lenses were all right, so the test was going ahead. Of course, I *knew* the lenses were OK, as I was willing to claim after the success of Trinity! I have excellent control of my nerves, so no one at the observation site knew of any reservations I might have had.

The observation point for us drones at the test was a hill called Compaña (sometimes spelled "Compania" so we Anglos could say it right), which was twenty miles away from the tower holding the gadget. Some fifty years later I returned to the site with my son John, who was working on a series of paintings of S-Site and the work we did there. We were treated to a personally guided tour of the site, and our guide told me that the name Compaña was on no map he had. I was happy to try to find it again, despite having been there only once and in the dark. I drove along the base of the several connected hills until the mountains forty miles across the desert looked just as I remembered them on that early morning long ago. I hadn't been eager to revisit this spot because of its association with my concerns about nuclear weapons and their use, but I dutifully identified Compaña Hill—or a hill near it!

There was a considerable wait before the test. There was no communication between Compaña and the tower. Twenty miles away, we did not know there was a weather question. The shot could not be taken if a wind of any strength was blowing toward habitation, for the fallout would be dangerous. Special teams had been stationed at nearby villages in case evacuations had to be ordered. Finally the meteorologist declared conditions acceptable, and Ernie Titterton started the first ever recorded countdown (they have become another cliché).

Suddenly, we saw it. The blast was brilliant against the predawn sky and distant black hills. A ball of fire rose in the cool air toward the tropopause. It took a couple of minutes for the pressure wave to reach us, and it was a satisfactory blast. I knew that Fermi would be dropping bits of paper at base camp, and I hoped they were suitably blown aside to give him an estimate of the yield. I had seen him try this elementary physics scheme before. The mark of a great physicist is to use what comes to hand, even for something so important. As any high school physics student should be able to tell, an object dropped from the hand will fall straight down with the usual acceleration due to gravity, but if there is a sideways constant force on the falling object, it will fall in a parabola away from the straight path in the direction of the force. Given the height of the fall and the extent of the lateral displacement of the object, one can calculate the force. In Fermi's experiment, the objects were bits of paper, because the lateral force would come from the wind caused by the nuclear blast some ten miles away. He was closer to the tower than we were, but working back from the wind force—an uncertain enough parameter—at base camp to the yield of the blast at the Trinity tower would hardly be definitive! But Fermi was irrepressible.

There was no need at twenty miles for the sunscreen and dark glasses that Teller sported, but it was clear that a major nuclear explosion had taken place. We could see few details of the explosion at our remove. I had borrowed a pair of field glasses, so I saw a little more than with unaided eyes. I watched the rise of the ball of fire, as I was to call it later when I studied the phenomena of a nuclear blast. The first "mushroom" cloud ever was formed quickly and rose rapidly into the sky. I noticed that at a certain point (I didn't know how high—perhaps fifteen to twenty thousand feet above the desert) a small cloud layer formed over the top of the rising cloud. I later had to figure out for the Bikini Tests what caused this unexpected phenomenon. The explosion cloud seemed to change colors as it rose, moving more or less through the spectrum, but how much of that was intrinsic and how much due strengthening sun light, I didn't know.

One of the most impressive effects of the blast was something I saw only a few days later, when photographs were available. Two balloons carrying instruments had been tethered near the tower. The photos showed that the tethers and the balloons had been vaporized by the expanding heat sphere from the explosion! I think if someone had asked me before the test what would happen to those balloons, I would have predicted they'd be vaporized, but still, when it actually occurred, I was impressed and gratified. Later I would calculate all the phenomena attendant to a nuclear explosion. It would seem that having such an understanding of a phenomenon in exhaustive detail would efface the wonder of observing it, but I have found the opposite to be the case. It usually surprises me — pleasurably — when I discover that something I've worked out in theory actually works out in the real world. To paraphrase Einstein, the greatest puzzle about the universe is that we can understand it!

two

Bikini

The Use of the Bombs

The relief among those present at the Trinity Test was almost palpable. Our gadget worked! The efforts of dozens of brilliant scientists and hundreds of able assistants over a period of nearly two years, working together in an unprecedented union of basic science and immediate practical application, had produced perhaps the most complicated "experiment" in history and brought it to a successful conclusion. Still, there was no cheering, no backslapping or congratulatory hugging as at the conclusion of a sports event. There was a bit of chatter and then quiet. I assume the thoughts of the others were like mine: How was the bomb going to be used? What would the existence of nuclear weapons mean for the future? As the bomb's creators, what was our responsibility in that future?

The United States was now in sole possession of the most powerful weapon in the history of warfare. The change of scale was so large that it had produced a change of character, a bomb of another species altogether. Even the thousand-pound chemical bombs already in use were no preparation for a nuclear device. Plus we had introduced another deadly feature—radioactivity, direct and from fallout—of which we would become increasingly cognizant in the months ahead. One plane (albeit a B-29) carrying just one of these bombs could produce nearly insupportable—and lingering—damage to the structures and population of any city in the world.

In July 1945 we had two of the new weapons, one Little Boy and one Fat Man. Obviously my fellow SEDs and I had no voice in the decision to use them, but we assumed that the political and military authorities would listen to the advice of our leaders. I don't think we were unworldly, but we were certainly naïve! I do not know what advice Oppenheimer gave President Truman, but ours would have been (without exception in my circle) to demonstrate the weapon to the Japanese military and invite surrender before it was used on targets in Japan. By the time we had working nuclear weapons, the Nazis had already been defeated. Whether we would have used them to shorten the war and save casualties in Europe I do not know, but the charge that our use of them on Japan and not Europe was racist is nonsense. In fact, the idea of staging a demonstration of the bomb for the Japanese high command had been considered and rejected because of the possibility that such a test might fail. The Trinity Test had been a major success, but that was no guarantee that the next gadget we dropped would not be a dud.

Another idea my friends and I considered was a drop in Tokyo harbor. If it worked, we could inform the Japanese what had happened, and minimal life would have been lost. If it did not work, we could try again—but it would take a while to make another bomb. I don't know at what rate Pu_{239} was being delivered to Los Alamos in the summer of 1945, but it would have taken my crew a few weeks to cast another set of usable lenses. We had too few molds.

Meanwhile, Allied military personnel were being killed and Japanese cities were being firebombed with significant civilian casualties. Any alternative to an invasion of Japan should have been tried, of course. We knew from the casualties we had suffered at Guadalcanal, Okinawa, Tarawa, and Iwo Jima, to name just four invasion battles, what we would face in Japan. (The fate of Allied prisoners of war was certainly in question as well.) War plans first drafted in the 1930s and updated regularly, anticipating a conflict with Japan, estimated a million Allied casualties (and five million Japanese) for Operation Olympic, the invasion of Kyushu, and even more for Operation Coronet, the invasion of the Japanese main island, Honshu. I do not fault Truman's decision to use the bombs, for he was accountable for every Allied casualty he had a means to prevent. I had no such responsibility. I just wish he — or we — had found a way to use them to stop the war immediately without making those of us who worked on them accessory to several hundred thousand deaths — and scarring wounds to thousands more — in Hiroshima and Nagasaki. I do not know about my friends, but I have never for a moment forgotten that responsibility.

Pressure Off

The pressure at the casting building at S-Site relaxed a little when our last bomb had been successfully exploded over Nagasaki. We were still casting lenses, but since the war was expected to end momentarily, the urgency of our wartime effort was gone. No one thought the Japanese could continue after two of their cities had been demolished with just two bombs — and they, of course, did not know that we had no others in reserve.

I asked for a three-day pass and went to supply to get C rations for a much-needed hike into the hills. I've never had much trouble working under pressure, but I react to it by getting tired. Wearing myself out physically seemed a good way to unwind. The problems of the bomb would disappear in the moment-to-moment activity of keeping track of where I was, finding water, making a comfortable camp, and observing the flora and fauna of the marvelous country surrounding Los Alamos. (I'm a better molecular biologist than taxonomist, but I recognize a number of species.) I still had the pack, shelter half, and other gear I had been issued when I left Camp Seibert, even though I was going back to college then and not into combat. (Regulations had to be observed!) So I rolled up a blanket and the shelter half, packed in the C rations, made my pack, and took off from Townsite.

I hiked over toward Bandelier and stopped one night just at dusk at a place that looked good for camping. I built a fire to warm my rations and settled down to sleep. That was when I noticed several pairs of eyes ringing my blanket. I thought they were probably mountain lions, which shows my ignorance then, since mountain lions do not travel in packs and would have avoided my fire. Quickly I built up the fire to keep the puma away and then in the brighter light saw that my visitors were cows! This was a puzzle: I knew the Lab had taken over some ranches when it was built (Anchor Ranch was still a name we used), but the cows had all been rounded up, hadn't they? Next morning the puzzle was solved. I had bedded down in an old corral, and some now feral cows that obviously had not been rounded up had returned to familiar haunts to see who had built a fire in their old place.

War's End

The Japanese surrender was signed on the USS *Missouri* on September 2nd, 1945, six years and one day after Hitler began the war by invading Poland on my sixteenth birthday.

Life on the Hill continued, though immediately after the armistice an exodus from Los Alamos began. The historic international gathering of physicists that had worked on the Manhattan Project began to disband, as they were anxious to return to their home universities.

We youngsters, SEDs and civilians, wanted to take advantage of their presence for as long as possible before they departed, so in the fall of 1945 we organized what we called Los Alamos University, a series of courses taught by these eminent men. Each of us was assigned to take careful notes for one course and write them up for everyone. I volunteered for Nuclear Physics. (It was during this time that I was introduced to the Breit-Wigner resonant model of nuclear reactions, which would influence the rest of my career.) What a faculty it was: Victor (Vicky) Weisskopf, John Manley, Norris Bradbury, Julian Schwinger. I'm sure I've forgotten someone. Schwinger had been at MIT during the war, working on radar, but came out to Los Alamos afterward as a consultant. His lectures were the most difficult to record. His board style was to write material with his right hand and follow with an eraser in his left! Some of what went into the "published" notes was, in Schwinger's case, reconstructed partially from memory. I have been told that these notes served as a textbook for nuclear physics courses for several years after the war, until a really good text (such as the book by Blatt and Weisskopf) came along. Leonard Schiff taught the course on Quantum Mechanics (which I attended, though I wasn't responsible for notes), and the notes for his course became his fine book on the subject. Geoffrey Chew taught Electromagnetic Theory, and Enrico Fermi gave Neutron Physics. It was in this course, I believe, that Fermi, writing a long series of equations on the blackboard, said, "It is obvious that—" and paused after writing down what was "obvious." He then turned to a side board, filled it with equations as well, and finished by saying, "Yes, it is obvious." I wasn't there, but one of my colleagues told me the story.

I never had time to go to parties during the war effort and so was not really a member of the social group that held them. But on New Year's Eve 1945 I was ready for a party and was invited to one at Fuller Lodge (the main building of the boys' school, which is still standing, if somewhat modified). The drink of the evening was Tech Area punch, a mainstay, I was told, of the wartime parties. (Prewar alcoholic drinks were still not in good supply.) Although it could be mixed in a bowl like real punch, which usually has five ingredients, my introduction to Tech Area punch involved the sip-and-fill method of mixing, with Coca Cola and lab alcohol (roughly 200 proof) as the only ingredients. It was simple: if you drank a good swallow from the bottle of coke and then refilled it with lab alcohol, you had a subcritical drink. Replacing two swallows made the drink critical, and with three swallows removed before refilling, it went supercritical! Since I had had no alcohol but 3.2 beer for many months and was not really a tippler anyway, I stayed with subcritical drinks—and not many of those. After welcoming in the new year, I started back to the barracks and a few yards from Fuller came across one of my mates sleeping peacefully in the gutter! With some effort and the fireman's carry I learned in Scouts, I got him into his bunk. When he was able to see through his hangover the next morning, he said he had no recollection of most of the night before, and after hearing how he got to the barracks, he decided that he might have died of exposure—despite the amount of alcohol in his blood-stream—if I hadn't stumbled over him.

Poker was the card game of choice among the SEDs, especially just after payday. My grandmother had taught me bridge (to make a fourth in the cutthroat game at ten cents a corner she played with her friends), gin, and poker, as well as several games of solitaire. Although I've played my share of cards, I'm not much of an enthusiast, and today I play only when my grandchildren demand it. (Once when I was about fifteen, I played with my uncle and his salesmen friends, thinking it was ten cents a corner as with Mama. When we settled up, I found it was one cent a point! My uncle staked me, of course, and fortunately for him, we won quite a bit.) Even so, I decided to play in a couple of the SED poker games and was soon asked if I would take a stake for half my winnings. I thought this bizarre, but there was no reason to risk my money if people wanted me to risk theirs. So I took subscriptions, and the deal was that after thirty hours of play, my back-ers would share half the winnings—and all the losses! I learned during those games why the Earl of Sandwich invented the eponymous food we have all enjoyed since. I don't think my backers ever lost, but I have not played cards for money since 1946.

We played baseball occasionally at S-Site during the war, and after-ward we played at lunchtime regularly. We also played football some-times, and I still have scars on the bridge of my nose from having my glasses broken by a shoulder to the head. We had no equipment, and so we usually played flag or two-handed touch. Many of the younger European scientists, refugees from Fascism and the British contingent, played soccer, and I joined in. I think Bill Rarita, from New York, and I were the only Americans playing. I usually got put at fullback because

my shooting wasn't too accurate. John Kemeny, a Hungarian originally, I think (he was a mathematician who went to Dartmouth later), explained my value to the team of more experienced players. "Mac," he said, "you make up in viciousness what you lack in skill."

Still in the Army

Those of us who were in the armed forces were not free to leave when the war ended. The demobilization of the Army of the United States could only happen gradually, as troops were still needed for occupation duty, and the economy could not absorb millions of new job seekers all at once. I would not accumulate enough points for discharge before the spring of 1946.

With the war over, the facts of enlisted life were more evident. Inspections had never been called during the war, but now I suppose the captain in charge of the SEDs thought to remind us of our status as GIs. He called a barracks inspection for the next Saturday morning. Unlike some of my colleagues, who had simply put on a uniform when they were inducted, I had actually been through basic and advanced training in army units preparing for deployment, so I had had a little experience with inspections. I planned the "GI party" for our barracks. I knew water was scarce on the Hill, and that we were lower than the civilian residences—particularly "Bath Tub Row," the houses that had been the quarters of the boys' schoolmasters and were now occupied by the leaders of the Lab, including Oppie. So we did a real cleanup, even scrubbing the floor with mops I moonlight requisitioned from the janitors. We got most of the footlockers packed according to the handbook and then set every shower, toilet, and washbasin to running all night. Come the inspection, we turned off all the water, got a compliment for the condition of our barracks—and never had another inspection! The civilians complained of no water that morning, and it was soon decided that the heroic efforts at cleanup had depleted the water supply too drastically.

Chapter Two

One Saturday afternoon a lieutenant from the PED, the engineers who built and maintained the Lab, came to the barracks to ask for volunteers. There was a fire in the forest north of the Lab, he said, and he wanted to take us out to fight it. I wondered why he hadn't gotten his own men—who might actually know how to fight a forest fire. He probably thought we SEDs didn't do enough physical labor as it was. I thought it seemed like an interesting way to spend a Saturday afternoon and so did eight or ten of my barracks mates. We were asked if anyone could drive a six-by army truck, and I said I could. (After all, I had driven explosives from Albuquerque, so I must be an expert driver!)

We started out toward the hills behind the Lab and soon were on a one-track dirt road among the trees, with the lieutenant leading the way in a jeep. A fire was yet to be seen. Soon we started down a long hill. I shifted down and gently applied the brakes. The way was bumpy, and I was trying not to bang the GIs around too much in the back of the truck. Just as the incline steepened, the brakes failed, and we began to pick up speed—as well as calls from the rear to slow down! Quickly I downshifted as far as the gearbox allowed and found that the gearshift would not stay in place. I told the GI next to me to hold it in; I needed both hands to steer this bucking vehicle, since the wheel was whipping back and forth as we slammed over bumps and ruts in the road. My seatmate was obviously terrified we were going to crash (as was I!), and he responded by releasing the gearshift and hollering. I told him that if he didn't hold on to it—and we survived the plunge down the road—I would kill him with my bare hands. He was only a little bigger than I was and apparently believed I meant what I said, as he held on to the gearshift (and stopped hollering). I was afraid I would not be able to keep the truck straight along the momentum vector we had acquired. I knew that if even a slight misalignment occurred, the truck would translate its forward momentum into a rotation about an axis across the road. While I might survive such a roll by holding tight to the wheel (seat belts were not around in those days), the GIs in the back surely would not. Thus whenever we were airborne, I had to make sure the wheels would be properly aligned when we landed. I have since landed small planes in stiff crosswinds with less difficulty than I had keeping that truck straight.

When we did reach the bottom, the GIs in back were shaken up pretty badly and the clutch was history, but no one was hurt. After some minutes the lieutenant came back to where we were stalled, and when apprised of the problem, took off to send back another truck. I was invited to ride with him and explain why I had ruined the clutch of a perfectly good truck! He was somewhat mollified when I explained and cursed the motor pool that had left the truck with no brakes. We never did find the fire that day. Unlike the devastating Cerro Grande fire of 2000, it may have simply gone out by itself.

The Bikini Tests

Now that the pressure was off, I wanted new challenges and a respite from the work at S-Site. Joe Hirschfelder and John Magee offered me a position with them in the Theoretical Division, as they were beginning a systematic study of bomb phenomenology—that is, all the phenomena connected with a nuclear bomb explosion. So three civilians from other casting groups—one for the development lab, one to see to transfer of new ideas to production, and one to be foreman of production—replaced me at S-Site, and I moved into Gamma Building next to Ashley Pond. (Some years later I learned that the pond—essentially a fire protection device in those days—was named for the headmaster of the old boys' school. His full name was Ashley Pond, to the great delight of the boys, who called the pond Ashley Pond Pond). Gamma housed most of T-division, which was headed by Hans Bethe. Joe Hirschfelder never did tell me how he decided to take me on. I suspect Paul Stein, who worked in T-division, had a hand in it. If so, he must have made a good pitch; I was between my sophomore and junior years in college as a physics major, although my electrical engineering education in the ASTP counted for something.

The army and navy were planning well-instrumented tests of the plutonium bomb in the South Pacific on an atoll called Bikini in the Marshall Islands. Major naval vessels would be moored in range of the bombs to see if they could withstand them and what kind of residual radioactivity would be produced in their metal fabric. Submarine torpedoes at Jutland in the First World War had raised questions about the usefulness of battleships, and Billy Mitchell (leading air commander in WWI and advocate of a separate Air Force) had shown them to be vulnerable to aircraft bombs. Now they would be tested against a nuclear weapon in an exercise officially dubbed Operation Crossroads. There would be two tests: Test Able, in which a Fat Man-type bomb would be detonated about five hundred feet above the surface of the water; and Test Baker, in which a similar device would be exploded by remote control at a depth of about one hundred feet underwater. (A third test, also underwater, was planned but never took place.)

Hirschfelder and Magee were part of the Los Alamos team assigned to help prepare for the tests. Both were theoretical chemists—Hirschfelder from the University of Wisconsin at Madison, and Magee from Notre Dame—who had chased fallout for the Trinity Test. Hirschfelder had been in charge of interior ballistics for the gun bomb (Little Boy) design and for blast studies for the Trinity Test. He was, therefore, ideally placed to study the phenomena related to a nuclear explosion. Our job specifically was to catalogue and assess the phenomena of a nuclear explosion for the people who would be determining where to place instruments to measure the blast, radiation, and fallout. To get the maximum information return on their effort

Cross Section of Assembled Bomb

(proportions are arbitrary. because of lens shape, no such cross section can actually be cut)

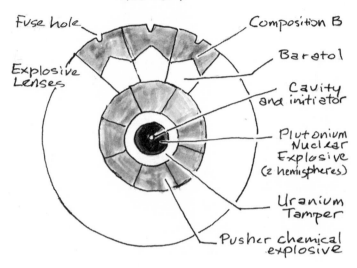

Fuse hole

Explosive Lenses

Composition B

Baratol

Cavity and initiator

Plutonium Nuclear Explosive (2 hemispheres)

Uranium Tamper

Pusher chemical explosive

they needed to have some idea of what to expect. My first task was to work out the details of the bomb designs beyond what I had already figured out during my tenure at S-Site. Any nuclear device needs an initiator to produce neutrons to start the chain reaction. Then subcritical pieces of nuclear explosive need to be configured to receive the initiator neutrons. A heavy tamper must surround the explosive to slow the disruption of the core and reflect neutrons back in to the active region. We used ordinary uranium (U238) for the tamper, which is denser than lead and can also enhance the explosion. (It also produces some unstable transuranic elements by neutron bombardment. In the environment of a nuclear explosion, rare events may become more common.) In the Fat Man design, a spherical shell of solid, fast chemical explosive surrounded the tamper to enhance the pressure of the blast wave, which was focused on the core assembly by a spherical shell of explosive lenses (those lenses that I knew so well!) in the outermost part. The implosion wave would crush the core assembly to start the initiator and increase the density of the tamper and plutonium so that a supercritical state would occur. Then a very rapid chain reaction could follow, using up as much of the nuclear explosive as possible before the core was disrupted. Detonators were set to initiate explosions simultaneously (within about a microsecond in 1946) in the thirty-two lenses. A steel shell enclosed the active parts. For a deliverable bomb, there was an external cover with fins to guide the dropped device. All of these components (which have analogs in Little Boy in a much simpler configuration) can become radioactive and contribute to the budget of radiation with which we had to deal.* With exhaustive study we could predict from the design of the bomb what would be the intensity of the shock wave, the anatomy of the explosion, and the composition of the explosive debris. Hirschfelder, Magee, and I were not the first to study the things that occur when a nuclear bomb explodes, but we may have made the first coherent attempt to do so. Interestingly, what we were doing was the converse of what is done today for monitoring nuclear explosions anywhere in the world. When a nuclear test or operational use of a nuclear device is detected, a thorough analysis of radioactive fallout can reveal the components of a bomb and ultimately identify who built the exploded weapon (but not, of course, who used it). Our work was crude compared to what is possible today, but it was a start.

The Shock Wave

For a number of the phenomena, we could use existing studies and fit them to our cases. First to emanate from the explosion is the shock, or blast, wave, which is familiar from experience with any explosion, although it is much more intense in a nuclear one. Shock wave effects

---

*Mockups of both Little Boy and Fat Man may be seen at the Bradbury Science Museum at Los Alamos and at the Atomic Museum in Old Town, in southwest Albuquerque.

had been well studied. In fact, the blast wave for the plutonium bomb had been modeled by an early digital computer. A series of calculations was run by a program wired into a plug board that controlled an IBM 602a, an electrical relay accounting machine that had been converted to a differential analyzer by Nick Metropolis, one of the bright young men Oppenheimer had brought from Berkeley. Nick understood (I think Karl Gauss had first pointed it out) that solving a differential equation step-by-step required no mathematics beyond arithmetic—if you knew what you were doing. (Later I would use an IBM 602a at Yale to solve the Schrödinger equation for nucleon-nucleon scattering, before we got a computer center.) Sometimes we assumed different bomb yields—the kilotonnage TNT equivalent—and calculated the effects as a function of yield. Punched cards could be prepared to input initial conditions—such as the bomb yield in tons of TNT equivalent— and slowly, slowly, Nick's program would generate a numerical shock wave, which I could plot.

Such a pressure wave produces an overpressure (that is, over normal atmospheric pressure) followed by an underpressure. On average, the wave moves at the velocity of sound for the pertaining atmospheric pressure, temperature, and humidity. Thus we could estimate the damage done by the blast at any point away from the center of the explosion. Assuming the wave expanded spherically, its intensity would fall off with distance as the square. If the bomb was set off near the ground, ground reflection of the wave would have to be figured in, as was the case for the Trinity Test. (The shock wave didn't matter too much at Trinity, since there was nothing nearby to be blown over.)

I discovered an interesting effect that had not been evident at Trinity, although after the fact I was able to find it in some of the photographs. In our library we had a whole collection of photographs of explosions, intended or accidental, including one of an ammunition ship exploding in a harbor. The photo captured an obscuring cloud enveloping the ship but provided no explanation. I knew that in air saturated with water vapor, the water vapor can condense into cloud if the air is cooled below the dew point—as happens near the ground at night to make dew. Any gas suddenly expanded by lowering the pressure may cool below the dew point if the humidity is high enough.

I reviewed the work of C. T. R. Wilson, a Scottish physicist at Cambridge, who had developed a chamber within which he could reproduce the conditions for cloud formation. Wilson's interest was in atmospheric physics (he explained the charge balance in the earth-air system), but he realized that his cloud chamber could be made to reveal the paths of charged nuclear particles. If a volume of air containing a volatile liquid (water, if you're studying atmospheric clouds) is rendered supersaturated by suddenly increasing the volume (done by pulling out a piston in the cloud chamber) with attendant cooling, it is ready to condense. If ions pass through the air, they leave droplets of condensed vapor along their paths. Wilson shared the Nobel Prize in 1927 for this discovery.

I reasoned that if the air surrounding a nuclear explosion had enough moisture, the underpressure part of the passing shock wave would produce a supersaturated condition, and the water vapor would condense on whatever particles—dust on land, salt crystals on sea— were around, just as the droplets formed on the ions of Wilson's cloud chamber. Thus we called this the "cloud chamber effect"; its signature for ocean explosions was a white cloud enveloping the explosion for a short time after it occurred.

We expected nearly saturated air all the time at Bikini, so the "cloud chamber effect" would be significant. My job was to calculate the thickness and duration of the expected cloud as a function of the humidity and the size of the nuclear explosion. I tabulated the assumptions I had made about those variables and produced an array of possible clouds. Much to my delight, the pictures of the sea level shot at Bikini amply show the expanding cloud. I never knew whether anyone actually checked my table, but the experimentalists at least had some idea of what to expect.

Why had no cloud chamber effect been anticipated at Trinity? The answer is that the New Mexico humidity is rarely high enough to allow a shock wave—even from a nuclear bomb—to produce supersaturation in the atmosphere. Yet when I reexamined the photographs of the Trinity explosion, I found the effect! As the shock wave expanded through the atmosphere high above the desert floor, it traveled through some layers that did indeed contain sufficient water vapor and were cold enough to become supersaturated. Thus the cloud formed as rings, or partial rings—exactly what you would expect of a spherical shell passing through a plane.

I could find little in the available literature on the electromagnetic pulse that a radiation wave would produce, so I had to become familiar with ion recombination to calculate that phenomenon. The idea was that radiation passing through the atmosphere would dissociate the air molecules into ions, and that these molecules would then come back together again, giving off some of their recombination energy as radio waves. I don't think anyone bothered to measure the actual effects of the electromagnetic pulse at Bikini to check my calculations, but the disruption of radio communication—for which I estimated duration as a function of yield—seemed to more or less verify them. One of the most far-reaching effects of a massive nuclear attack would be on electronic communications—something to consider in today's electronically interconnected world.

The Heat Wave

Intense heat is another of the phenomena we studied. It occurs in any explosion and can be magnified to devastating levels when incendiary materials are plentiful. The incendiary bombing—and its incident fire storms—of Hamburg, Tokyo, and other cities during the war had revealed the destructive capability of a heat wave when old wood (Hamburg) or light construction (Tokyo) is present. As with the shock wave, the heat wave is much more intense with nuclear weapons. I had noticed in the pictures I saw after the Trinity Test that the instrument balloons tethered near the tower had been vaporized—tethers and all—as the heat wave expanded. The expectation at Bikini, however, was that most of the ships would withstand the heat from the explosion. Knowing the temperature at a given point within the explosion and the fuel load of the target, we could estimate the destruction each ship would undergo. We anticipated that fires would start in the flammable material of the ships (there's actually quite a lot), but that the structures themselves would be, at most, distorted and not destroyed. (This projection says nothing, of course, about the effects of the blast wave.)

All observers at Bikini would be evacuated to safe distances, so we were less concerned about the hazard presented to them by heat than by some of the other products of the explosion. Although observers reported that they felt "as if a furnace door had been opened" when they saw the ball of fire at Bikini, this was more of a psychological response than an objective one. Had our calculations been applied to Hiroshima or Nagasaki, on the other hand, they would have forecast tremendous heat damage to all human life and buildings within a considerable radius of the impact points. Indeed, photographs from Japan later showed that the fires induced by heat from the nuclear explosions at Hiroshima and Nagasaki (initially near the temperature of the sun) caused overwhelming loss of life and near-total destruction of the nearby buildings and infrastructure in moments.

In 1946 we were dealing with one bomb at a time, but years after the war scientists conducted theoretical studies of what would happen in a full-blown nuclear war, with the detonation of thousands of nuclear bombs. They projected that immense fires would be ignited in the forests of Sweden, leading Carl Sagan and others to forecast that the resulting dense clouds of smoke would obscure the sun so thoroughly that the earth would be plunged into a deadly "nuclear winter." While their atmospheric models have been questioned by some, I believe, with Sagan, that this is one experiment the world cannot afford to try!

Fallout

Radioactive fallout was one of our biggest concerns, and we tried to study it in a little more detail than had been done just before the Trinity Test. At the explosion point of a surface blast, the heat causes the air to expand and become less dense than the surrounding atmosphere. Essentially a hot air balloon without an envelope is formed, and as it rises it takes much radioactive debris with it. The rate of rise of this "ball of fire" can be calculated if the initial and ambient temperatures are known. In general, if there is no wind, the ball will rise to the top of the weather region of the atmosphere, called the troposphere. If there is a wind, some of the debris will be carried along with it and will eventually settle to the ground—this is the fallout, of which much has been written. It is highly radioactive, consisting of the fission fragments from the nuclear explosive and the material of the bomb's structure—all of which has been vaporized, and, like the surface material sucked up into the ball of fire (if the explosion is on the ground), made radioactive by neutron-induced nuclear reactions.

It was important for any planes involved in the Bikini Tests to avoid the ball of fire and the fallout it carried, so I had to learn some atmospheric physics and meteorology. I broke the problem down into two parts: the rise of the ball of fire and the debris it carried, and the drift of material from the initial site at both low and high altitudes.

I learned that prevailing equatorial air currents might help carry radioactive debris at low altitudes toward downwind sites, but we would not have much data on local winds. We would just advise that, as at Trinity, the test be initiated when winds were calm. I was more concerned with the fate of the higher-level debris, which might be injected into air circulation at any level. I thought to learn about the higher-altitude drift of debris from volcanoes, preferably in the sea, and found the ideal case in the explosion of Krakatoa, a volcanic island that had disappeared in an enormous eruption in 1883. I read that ashes and pumice from Krakatoa were deposited a hundred miles away, and that clouds obscured the sun at a hundred and fifty miles. The finer particles from the eruption rose to the top of the atmosphere and caused brilliant sunrises and sunsets around the world. As powerful as they are, nuclear bomb explosions do not approach the intensity of this mighty explosion, which had turned the island mountain into a suboceanic cavity, but the reports I read gave me the basic information I needed.

Our calculations had indicated that the ball of fire from a nuclear explosion would stop rising when it met the stratospheric temperature inversion at the base of the stratosphere (perhaps 35,000 to 40,000 feet). This is where the temperature of the atmosphere begins to increase with height rather than decrease as it does down in the troposphere (except where there are local inversions). If there was significant wind, the radioactive debris would drift as it rose. If it got into the jet stream, it could go around the world, diminishing in intensity as it spread out. The jet stream had recently been detected by bombers attacking Italy, for it could cause extreme drifts off course or slow ground speed so that fuel would be expended at an unacceptable rate. I didn't know about it at the time, and no one knew its path anyway, but I knew that the fall-out from Krakatoa was carried almost around the world in a swath fifteen degrees of latitude wide. We could assume that radioactive fallout from nuclear bomb explosions would behave the same way. Of course, experience has since verified this expectation.

Despite the volcanic data, we could model the stratospheric distribution of fallout from the Bikini Tests only very crudely and advised that it would get dispersed well enough and be of low enough density to be harmless to people downwind if they were far enough away. (The native peoples had been removed to other islands, whether they wanted to go or not, and have only recently begun to return.) We made reasonable assumptions based on the best information we could get, limiting our scope to the task at hand. Since then studies have been done with better atmospheric models and much better data, exposing fallout for the major hazard it is. Now, of course, it is seen as one of the gravest dangers of nuclear war with the potential for drastically reducing or even eliminating the human population of the earth. I can only hope that this apocalyptic possibility will continue to deter sane political leaders from starting such a war.

Marriage

By early March of 1946, I had finally earned enough points to gain a discharge from the army and was sent to Fort Bliss, near El Paso, to be processed out. I respectfully declined an invitation to join the active reserves and walked away a civilian again after a few days more than three years in the AUS. Hirschfelder invited me to come back to Los Alamos as a civilian and continue my work on the phenomenology for the Bikini Tests. My fiancée, Mary Muska, seemed to think that was a good way to start married life before going back to school in the fall, but first I had to journey east to Pittsburgh to get married.

I have no idea what difficulty my timetable presented for the wedding preparations in the Muska household, but it certainly caused some trouble for Mary's priest. We needed dispensation to get married in the church, for I was Protestant, and it was Easter season. The Pittsburgh diocese was not noted for its speed in the best of times. Finally, the young priest, Father Brennan, went downtown to Pittsburgh from Coraopolis, some ten miles up the Ohio River, and picked up the dispensation himself. His instructions to me were succinct: Mary would be allowed to practice her religion and raise our children as Catholic. I had no problem with this. Father Brennan probably got into trouble with his dour senior in the church, but we were all legal according to canon law. Mary's mother had to sign for her to get a license, since she was a few weeks shy of twenty-one. The next day we headed back to Chicago to visit my parents, and after a day or so, boarded the Santa Fe Chief for Lamy.

We stayed in Santa Fe until Mary got cleared by Los Alamos security. Wartime procedures were still in place, and wives of lower echelon personnel like me had to have a job in the Lab and had to be cleared to get one. Mary's parents told us of neighbors, schoolteachers, and others in Coraopolis asking just whom had she married that all these FBI types were in town asking questions about her!

When Mary's clearance came through, we went up the Hill to reside in a married couples' dormitory. We had a small room with army bunks I soon redesigned as a double, an electric hot plate, and a refrigerator on each floor for general use. We had a lovely lady who cleaned the room regularly. I suppose this was better than providing cleaning supplies for everyone—anyway, it gave her a job! We ate most meals in the mess hall nearby. I was accustomed to army food, and Mary had little choice given our facilities.

Mary was offered a job with Lieutenant Ernestine Savelle in the travel office, and I went back to work in Gamma Building. Mary met many of the "big names" on the Hill, sometimes arranging their travel back to their postwar lives. One of these was a visitor named Gregory Breit, who appeared one day just before noon seeking to have his transport back to Madison arranged. The office personnel had recently been warned by the major in charge that lunch hour began promptly at 12:00 P.M., so Mary excused herself and told Breit she would be back at one o'clock. When she returned, there he was, patiently waiting for her help, which was promptly given. Neither of us knew then that our association with him would be close for the next forty years.

Security was still on a wartime footing. The work of the Lab was still classified. The nature of my work for the Bikini Tests now qualified me for a white all-area badge, giving me access to anyone, anywhere, and to any document. With her restricted badge, Mary could not visit me in my office. Ernestine fixed that by getting her a yellow (absolutely no secrets divulged to a yellow badge!) all-area janitor's badge. Some may have wondered at her dress for janitorial work, but the badge got her in to see me. On the intercom for Gamma Building, where all kinds of fake and silly announcements were made, the most frequent call was for "J. J. Gutierrez," head janitor!

Mary and I had a bit of a social life. A couple in our dormitory liked to play pinochle. The man was in Supply, and their bed hardly sat on the floor from all the supplies he had "requisitioned." I don't know how he expected to ship the stuff back home when he got out, but I've heard of much larger projects of GIs in Europe, so I suppose he managed. Sometimes our pinochle games got a little tense. Mary and I partnered against him and his wife, and he was prone to blame her for any difficulty their partnership had. Once she paid him back in one hand by playing every card to our strength!

People our age made home brew (better than the 3.2 beer at the PX!) and had parties at their houses on the weekends. I recall one where at 10:00 p.m., Mary and I decided to get all the blankets we could find and cover the partiers, including the hosts. The home brew had taken its toll: everyone but us was asleep—on couches, the floor, beds, whatever! We went back to our barracks and were in fairly good shape the next day when we went over to help clean up. The others had gone home, but the hosts had major hangovers, so we did most of the work.

The most memorable party we attended was at the home of Louis Rosen. Louis was a senior physicist at the Lab who would later become an adviser on nuclear disarmament both to our government and the Russians. (He made me a consultant when I returned to New Mexico.) It was an indication of the intellectual tone of society at Los Alamos during and after the war that the entertainment at the Rosens' party was to be a reading of *Taming of the Shrew*—and that enough copies were available from neighbors for each of us reading a part to have a copy of the play. Such dramatic reading had been a regular form of entertainment during the war, but I had never participated because I worked so far away.

My own central recollection of the party, however, began at dinner. Mrs. Rosen served a roast, and Louis was late—some Lab business detained him, of course. When we sat down, Mrs. Rosen presented me with the cutlery and asked me to carve! I never refuse a challenge, so I did my best, but it was a slightly traumatic experience! The guests were kind enough not to complain of my results. I have since studied the anatomy of all fowl and cuts of meat served at table and can carve any of them satisfactorily.

Radiation

The first Bikini Test, Test Able, was now scheduled to occur on June 30, 1946, so the pressure was on when I returned to work. The Navy was somewhat belatedly becoming concerned about radiation. We already had a study on our agenda in which we were looking at exactly how a nuclear device produces radiation and of what kinds.

A nuclear explosion begins with fission, and fission produces nuclear fragments that are neutron-rich isotopes, meaning they have too many neutrons to be stable in the binding environment of the nucleus in which they find themselves. By a series of beta decays, the extra neutrons become protons, and the fragments reach stability in the table of isotopes. Six or seven electrons are given off in the process (neutrinos are too—but we needn't worry about them). A flood of neutrons is produced, and anything in its path that is susceptible to neutron activation—and most materials are—becomes radioactive.

Fermi had studied neutron activation in 1934, and others had subsequently found that the induced radiations are so characteristic that the process can be used as a tool for chemical analysis. The nuclei created by this neutron activation emit radiations ranging across the spectrum—alpha particles, protons, gamma rays, and, of course, the ever present neutrons and electrons. We knew something of the scattering length—the average distance between collisions—of these nuclear particles in air and hence could estimate their toxicity at given distances from a nuclear explosion if we assumed a particular explosion size (such as ten, fifteen, or twenty kilotons TNT equivalent). All the constituents of the bomb—even its casing—would become radioactive with different intensities from neutron activation and be driven around the blast area by the blast wave or rise with the ball of fire and the mushroom cloud.

Each of the various radiations emanating from a nuclear explosion is lethal in large enough doses. In fact, the mass, energy, and intensity of a given radiation can be summarized in a unit of damage to the human body. Penetrating radiation ionizes the material through which it passes, meaning that it creates a charge in any atom it encounters, and such an ionized atom ceases to function in living processes. With charged particles, the ionization is direct—one or more electrons in the shell structure is knocked away—but with the neutron, a collision with a proton produces a "knock on" charged particle that does the ionization. The charge-neutral neutron can travel a great distance through a material before it makes an ion, which needs to be taken into account in estimating the damage capability of a flood of these particles.

We knew from Hiroshima and Nagasaki that the radiation from a nuclear explosion in 1945 would be lethal enough to kill thousands of people in the target area, both from the initial blast and from the residual radiation. At Bikini, Naval personnel would be well away from the center of the explosion, but they would be boarding the tethered ships afterward to study the damage. It was up to us to help determine when it would be safe for them to do so. The body can recover from a limited amount of radiation damage, which is why we survive diagnostic X-rays or radiation therapy for cancer. Cosmic radiation beats down on our heads anytime we're outside, and some of us even seek out the ultraviolet component of the sun's radiation for tanning. But in 1946 it was not really known what the upper limit of a safe daily dose of radiation was, and we are still not sure there is a threshold below which there is no lasting damage. The people examining the ships after the Bikini blast would be equipped with the best radiation detectors we had then, but these would be of limited value since we couldn't really determine what were unacceptable levels of radiation. Still, we examined the various radioactive products of a nuclear explosion in some detail and estimated how far they would travel. We knew that once settled on a ship's deck, they would be dangerous until scrubbed off, and that some neutron activation of the ship's structure would occur. All we could do was estimate how long one might stay on deck with reasonable safety.

One of the radiations that warranted extended study was the gamma ray, high-energy photons emitted in the fission process and by irradiated material from the bomb. We knew that deadly gamma rays are deeply penetrating, but also that their penetration range is finite. The question was, what is the range? Both for the safety of personnel and instrumentation, we needed to figure it out. I got that assignment because physicists are supposed to know about such things, and my senior colleagues were theoretical chemists. As I was at least two years away from a baccalaureate in physics, it's not clear that their faith in me was well placed, but I went ahead.

I began to study Heitler's book on quantum electrodynamics, the bible on the subject then and for years after. I was interested to find that in his introduction Heitler thanked Klaus Fuchs for reading the manuscript. Conveniently enough, Fuchs had the office across the hall from me. I didn't hesitate to ask him about anything in Heitler I didn't understand. Nothing in my observation of Klaus Fuchs's behavior suggested his role in espionage. We were not close friends, but friendly acquaintances. I do not recall discussing politics at all. I was as distressed as anyone when his role in wartime espionage came out.

Soon I was able to catalogue the effects—Compton effect, pair production, and photoelectric effect,* for example—that gamma radiation passing through matter would undergo. These effects conveniently arrange themselves in energy order. The Compton effect is most important at higher energies, photoelectric at lower, pair production unimportant after the photon has lost enough energy, and so on. As the gamma ray moves through matter it collides with individual atoms and loses energy with each collision. The photon leaves each collision with an angular distribution that depends on its incident energy for that collision, and the three effects dominate in succession as the energy decreases.

The scheme I concocted was to average the photon intensities in a beam of photons over the angular distribution of their scattered directions when incident on a given target atom or molecule. Then I would take the angle for maximum intensity as the incident angle for

---

*Arthur Holly Compton showed that a photon scatters off an electron exactly like a particle (the Compton effect), with conservation of momentum and energy (so the photon loses energy to the scattered electron). He thus confirmed Einstein's postulation of the photon as a particle made some years earlier in discussing the photoelectric effect, where a photon is absorbed after it ejects an electron from a surface. A third way a photon can lose energy is by creating an electron-positron pair, with the mass and kinetic energy of the pair coming from the photon.

the next collision. The energy of the scattered photon could be similarly selected once the angle was chosen. I could add up the results and bias them according to the percentage of molecules of a particular type in the material traversed (nitrogen 78 percent and oxygen 21 percent for the atmosphere, for example). In this way I could estimate the energy of the photon at any distance from the source given its initial energy (the energy distribution of fission-induced gamma rays was known) and the kind of atoms it was encountering. I could then sum over the initial distribution and calculate the flux of gamma rays at any distance from the source until the gamma ray was finally absorbed by the matter traversed.

This method yielded sensible answers for the limited data we had to check, and Hirschfelder accepted it as our calculation for this phenomenon. (The Bikini calculations were for air, but it was little trouble to change the material.) After we dispersed to our respective universities, Hirschfelder, Magee, and I published the report I had prepared in *The*

Physical Review, the leading physics journal in America. Its title was "Penetration of Gamma Rays Through Thick Targets," and it dealt with Compton scattering, the most important effect in gamma ray penetration. On rereading the paper for the first time in some fifty-six years, I see we promised to publish the rest of my report—on pair creation and the photoelectric effect—in a subsequent paper. We never did. Joe and John went back to prewar careers, and I was about to start working on my PhD. But I was pleased with our paper of 1948, for it was my first. Published during my senior year at Yale, I understand it established the accepted method for calculating radiation shields for accelerator installations until faster computers made Monte Carlo calculations possible. *

As we worked through our calculations it became clear to us that the various radioactive products of a nuclear explosion—whether in the fallout or coming directly from the explosion—are uniquely inimical to life. They do not occur in a chemical explosion. The force of a nuclear explosion is many times greater than that of a conventional bomb, of course—the bomb exploded at Trinity was estimated at 18,000 tons of TNT equivalent—but its deadliness extended far beyond the force of the blast wave. We had known since Marie Curie studied radioactive species that the radiations were dangerous, but she dealt with fractions of a gram of radium or polonium, while our studies dealt with tons of radioactive material—from only one bomb! If thousands of bombs or warheads were exploded at one time in a nuclear war, the spread of radioactive debris over the earth could jeopardize the continued existence of life on earth. Nuclear weapons had not only raised the scale of devastation beyond what we could previously extract from explosives, they had given humankind the power of

---

*Monte Carlo calculations are based on the idea that one can get a reasonable approximation to the solution of an equation by randomly sampling its contributions. Stan Ulam invented the technique (with von Neumann) at Los Alamos. It needs fast computers.

utter self-destruction. One can hide from a blast wave, but there is no easily available place to hide from radiation. (The effort to build back-yard fallout shelters in the '50s was a response to this realization, but they wouldn't have worked. Eventually, one would have had to emerge to get supplies, and some of the residual radiation would still have been potent.)

Hiroshima and Nagasaki

By the end of 1945, we phenomenologists had been given pictures of Hiroshima and Nagasaki against which we could test our calculations. They were only marginally useful. A warship is far sturdier than a light frame building, or even a reinforced concrete one, so our estimates of ship damage at Bikini were considerably less than what we saw in the Japanese cities. But the photographs affected me. I have never been able to look at them, or, especially, the Signal Corps films of Hiroshima and Nagasaki taken after the occupation, in a purely objec-tive manner. I have seen photographs of Civil War dead on several bat-tlefields—the Sunken Road, Marye's Heights, the Wheat Field—and am saddened by the waste of life represented by the piles of bodies pro-duced by the minié balls from both sides. But these men were killed by each other in battle in a war—however irrational—that was waged at the level of personal conflict for its participants. The scenes at Hiroshima and Nagasaki were wholly different. They were not battle-fields, and the bodies were not soldiers. They were civilians annihilat-ed while engaged in the various acts of daily life.

Here had been living people whose bodies now littered the streets, homes, and working buildings now in pieces scattered about; walls that had been erect moments before Enola Gay dropped her load were now bent over as if tired of standing. One man had been caught running across a bridge when the heat wave came; his silhouette was etched on the walkway where his vaporized body shielded the scorching of the structure for an instant. Pictures of the Hiroshima Maidens, young

women grossly disfigured from thermal burns, seen since the war have only increased my discomfort. Long after the war, others would succumb to radiation-induced cancers. This never happens with chemical explosions, no matter how powerful. I knew that there must also have been many people who lingered with radiation sickness and radiation burns, though the early (Japanese) pictures I had didn't show them. Perhaps no one but the young can be truly innocent in a war, but these people were not directly involved in the fighting, and they had no warning.

During the war the urgency of our task and the constant press against time excluded other thoughts from our minds—or mine, anyway. There was a sense that our efforts might help bring an end to the war, but our immediate goal was simply to get the bomb built. We faced enough problems in chemistry, physics, and especially applied physics to satisfy any of one us, and we focused on them with the single-minded intensity characteristic of all scientists. It was not until after the Trinity Test that we could stop and think about what we collectively had done. My GI colleagues and I had misgivings about how the bomb would be used, but these misgivings dealt with scenarios. (I would later learn that most other project workers shared this concern—and that General Groves had suppressed our opinions.) In working on bomb phenomenology for the Bikini Tests, I had to dissect the effects of real bombs on real cities.

Still, Japan had started the war, and at the time many (or most) Americans felt that its people got what they deserved for supporting their leaders. We were fighting a defensive war, which there was no question about supporting. Members of my family had been in the military in every major conflict in U.S. history—the Revolution, War of 1812, the Civil War (on both sides), and four during the twentieth century—and I am certainly no pacifist. Still, I don't always agree with the way our leaders initiate or conduct armed conflicts. The casualties in Hiroshima and Nagasaki still haunt me. And it was not difficult to visualize New York or Chicago or Washington in those pictures.

In 1946 we had a monopoly on the bomb, but we knew that the biggest secret about nuclear weapons was that they could be built. (Thanks to my sometime colleague Klaus Fuchs, that monopoly proved to be short lived indeed!) Any technological society could, in time, have its own nuclear arsenal. Up on the Hill in the rarified New Mexico air, the remaining cadre of scientists who built the bomb continued to study its effects, but we knew that as soon as the photographs from Japan were disseminated to the public, everyone would see the effects that had once been the privileged knowledge of those of us who could calculate them. Nuclear weapons would never again be under our sole control, but there was a growing conviction among us that they must never be used again. We hoped the pictures from Japan would help convince the world.

The Slotin Accident

By late spring of 1946, the people involved in Operation Crossroads had all gone out to the Marshall Islands, including Hirschfelder and Magee. The theoreticians were needed back at their computers, not on board ships, so I didn't go. I didn't mind since I was newly married, and separation so soon would have been difficult. I was to maintain the office and answer any queries that came from the South Pacific. Matters of our concern seemed to have gone well. I got no frantic calls for help—just an occasional routine enquiry.

There was one small disaster during the tests, which amply illustrated Alexander Pope's remark on a little learning being a dangerous thing. We lost some data when a naval officer was prohibited from boarding a ship after the blast because his radiation counter read too high. With his quasi-expertise, he had opened a window and thus was counting electrons (not too dangerous) rather than alphas (very dangerous). The Navy wouldn't allow civilians on the boats, and the officer had not been sufficiently educated in the use of his counters. But that was not a phenomenology problem.

The end of the war did not end the mission of Los Alamos, and work to improve the weapons we were making continued. One goal was to refine the size of the critical mass, for the less nuclear explosive that was needed, the less effort would be required to make enough for a bomb—or, eventually, for missile warheads. Several experiments had been devised to measure critical mass in nuclear charges of the size and shape that might be used in a weapon. One arrangement, called the Guillotine, allowed a subcritical piece of uranium to fall through a hole in another subcritical piece, producing a momentary critical mass during which time the neutron flux could be measured. In another, a bare "Godiva," a nearly critical mass, could be covered, or "dressed," piece by piece, with a material that reflected neutrons back into the charge where they would make more fissions. This was called putting a tamper around the charge, and if the charge was right, the approach to criticality could be measured when the tamper was added gradually. Or one could "Tickle the Dragon's Tail" by gradually bringing two subcritical hemispheres of plutonium closer together and again measure the approach to criticality. I had heard there was an accident during one of the critical experiments, but it had happened just before the war ended, and there was no noise made about it.

Toward the end of May, a tragedy put a damper on the whole Hill. A physicist named Louis Slotin, who as a Canadian may not have been able to go to the Bikini Tests, was continuing his work on critical mass and teaching other experimentalists how to do the tests. During the war, when time was the most critical element of our work, strict safety was sometimes replaced by the informed judgment of the person in charge (as I had done at S-Site), and Slotin may not have gotten out of wartime habits. As I was subsequently told, he was demonstrating approach to criticality with two hemispheres of plutonium. He placed one above the other, with rims touching on one side and a screwdriver separating the rims on the other. The approach to criticality was made by turning the screwdriver slowly toward its flat side, thus closing the gap between the hemispheres, and the neutron flux from the setup was measured continuously as the gap narrowed. This time the screwdriver slipped, the hemispheres came together, and a flood of neutrons came from the device. Instead of trying to shield himself from the deadly device, Slotin immediately put out his hand and broke up the arrangement, thus stopping the neutron flux. This saved his students—not from serious radiation poisoning, but from death.

I was not acquainted with Slotin before the accident and never saw him afterward, but I came to know him in a ghostly way. With many of the leading scientists away in the South Pacific, I was asked to estimate the dose Slotin had received. The neutrons had activated materials in his clothes and pockets, and the level could be measured. From this the maximum flux could be estimated and from that the dose he got. I assume anyone left on the Hill who could do the calculations did so, but I knew only my own results—something like twice lethal. Slotin was dying, and his family was called from Winnipeg.

Klaus Fuchs was also called on to look at aspects of the accident. He did some calculations to see what would have happened if Slotin had not broken up the arrangement of hemispheres after the screwdriver slipped. He concluded that the assembly would have heated enough to expand and go subcritical. Then it would have cooled enough to contract and go critical again, and so on. So there would have been an oscillating source of deadly neutrons in the laboratory, and the students would certainly not have survived. Slotin's move probably made no difference in his own case (we can't be sure because "dose" is a product of flux and exposure time, and he could have reduced exposure time by withdrawing instead of putting his hand in the flux), but he did not know that at the time.

I learned that, ironically, Harry Daghlian, who was killed in an unpublicized accident just after Nagasaki, had been Slotin's assistant, and that Slotin had stayed by his bedside as he died. (It took a month. Slotin lingered for nine days.) So far as I know, there were no other lethal accidents of any kind at Los Alamos to that time, but the death of even one, unknown colleague in a relatively close-knit group is distressing — "never send to know."* We knew that death from radiation is not pleasant, but few of us had actually seen it before these tragic cases at the Lab. My calculations for Bikini and the pictures from Japan now had a familiar face — Slotin's.

Endnote

I have never seen any written reports from the Bikini Tests, but I heard they went off satisfactorily. The bombs exploded as planned and the ships were heavily damaged, though not all of them sunk. No battleships have been built since the war, but the naval conflicts from 1939 on would have advised that without Operation Crossroads. The tests showed that a concentration of ships was vulnerable to destruction from a single bomb, but Pearl Harbor had already shown the danger of concentration of assets in a port. The underwater test would have shown ship vulnerability to nuclear torpedoes and the air tests to missiles that we now have. But all this was no longer my concern.

---

*John Donne's Meditation XVII begins with "No man is an island," and ends with "never send to know for whom the bell tolls; it tolls for thee." Both ideas are appropriate for the band of brothers created during the war at Los Alamos and the death of one of them.

three

Yale

Our Last Days at Los Alamos

With Hirschfelder and Magee away during the summer of 1946, Fred
Reines invited me to work with him on some neutron physics, and so
my physics "education" continued. (Reines would later earn a Nobel
Prize with Clyde Cowan for the experimental discovery of the electron
neutrino.) He suggested I look at the absorption of neutrons as they
pass through matter. It occurred to me to model their absorption as
one does light when it is absorbed in some media. With light one gives
the index of refraction an imaginary part—the index becomes complex. A complex number in mathematics is one that has two parts: one
an ordinary number we call "real"; and the other a real number multiplied by the square root of minus one, which we call "imaginary."
There is an "algebra" for complex numbers. Physical quantities, such
as the intensity of light passing through glass, must be real; when one
works with complex numbers, the result of calculating a physical
quantity can be smaller than if one had worked only with real numbers. This reduction represents absorption.

It had been shown that neutrons' behavior, like all elementary
particles, must be represented by a "wave equation"—an equation
suitable for describing the propagation of a wave. Erwin Schrödinger
had proposed such a wave equation for material particles; thus "wave
mechanics" would replace Newtonian mechanics at atomic dimensions. Schrödinger's formulation incorporated the ideas of Max
Planck, Einstein, Niels Bohr ("Uncle Nick" at Los Alamos, after his
code name, Nicholas Baker) and Louis de Broglie on the quantum

nature of energy, the quantum structure of atoms, and the wave characteristics of atomic particles. Interaction, such as a collision, between two particles could be represented in Schrödinger's scheme by a force potential called a "potential well" because attraction between two particles was a negative energy region like a well into which the particle could fall. I therefore decided to use a complex potential well in a Schrödinger wave equation to represent the absorption of neutrons by matter through which they passed. The real part would describe the regular interaction and the imaginary part the absorption. The well would be in my simple calculation what we called a "square well" (the friend of all theoretical physicists!), meaning that its radial profile in the wave equation would have straight sides and a finite depth.

I searched the literature and found in *Zeitschrift fur Physik*, as I recall, a note on a similar idea by a physicist named Masataka Mizushima. The note was brief and its author was not a native speaker of German, so it was not much help, but at least I knew that someone else had had a similar idea. Using my square well, I separated the Schrödinger equation into two real differential equations. It turned out that the solution had the form of the half-odd-order (i.e., 1/2, 3/2, 5/2, . . .) Bessel Functions, which led me to a book in which G. N. Watson, a British mathematician of the twentieth century, had described the properties of the solutions of the differential equation studied by the German astronomer F. W. Bessel in the early nineteenth century. I knew I could calculate the Bessel functions from the series in the book, but I thought someone must have published a table somewhere. Sure enough, the librarian found a set of tables that had been calculated during the depression as part of a WPA project created to employ some New York mathematicians. I happily calculated my cross sections—the chance of absorption as the neutron passed through the material—for Reines, adjusting the potential for the available data. The result was a fairly good model.

One morning I discovered that the security people had come into my office overnight and classified the tables "Secret"! Although this was insane, it didn't cause me much difficulty until the day a visiting mathematician came around to borrow them. I couldn't give them to him because his temporary clearance was not high enough. That was bad enough, but the final irony was that he was one of the authors of the tables! I solved the problem by letting him use the book in my office, no doubt violating all kinds of rules.

Reines apparently knew that Vicky Weisskopf was also thinking about a complex potential to model nuclear reactions. He asked me not to discuss our results with him, as we might wish to publish them if they could be declassified. That seemed odd to me: what could Weisskopf possibly need from an undergraduate who was not even a certified physics major yet? But Weisskopf already knew about my work, I learned, when we found ourselves together on the Chief bound for Chicago.

Back to School

At last, my work at Los Alamos was over, and Mary and I were going back to school in August of 1946. On the train, Weisskopf and I discussed the work I had done with Reines, and he told me a little about his ideas on the subject. (This was one of many times when Mary has complained of physics occupying too much of my attention when she was around!) I was flattered by Weisskopf's attention, as you can imagine, and so ignored Reines's request. As it turned out, Reines and I never tried to publish the work. I suspect it was not deep enough, but it couldn't be declassified then anyway. My indiscretion was rewarded a year or so later when Weisskopf visited Yale to gave a talk about his "cloudy crystal ball" model of nuclear reactions and acknowledged our train conversation in the early days of his work. I was still an undergraduate at this point (a senior), and my status among the graduate students increased considerably!

Mary and I visited my parents briefly when we got to Chicago, and then I set out for Madison, Wisconsin. She was to follow when I found housing. I had been accepted with junior status in the undergraduate physics program. We were going to the university there at the invitation of Gregory Breit, the customer Mary had put off to go to lunch with me! Breit had been among the very important physicists invited to the Hill when Oppenheimer hosted what amounted to a special meeting of the American Physical Society to celebrate the end of the war and the contributions physics had made to its successful outcome. I recall a session with four (or was it five?) Nobel Laureates in one row! Breit had been head of the Neutron Committee, the progenitor of the Manhattan project, but had resigned because of differences over security. Deciding that the development of the bomb was in good hands anyway, he went on to do very important war work on the proximity fuse and protecting ships against magnetic mines. Having accepted that I wasn't going to join him in theoretical chemistry, Joe Hirschfelder recommended me to Breit when he came to Los Alamos.

I first lived in a converted sheep barn on the agricultural campus of the university in Madison. So many ex-GIs were coming back to school on the GI Bill—as I was—that there was not enough housing for them. The room was comfortable enough, but the smell of sheep droppings had not been eradicated. I eventually got a room in a converted barn in a village outside Madison where there was bus service to town, and other students had cars for commuting. Mary then joined me and got a job at the university to supplement the GI Bill.

I was initially assigned to assist in the sophomore labs, but Breit soon took me on as a research assistant. I was taking the usual junior curriculum for physics majors, except that I had to make up a sophomore course in English I had missed in my checkered college career. It seemed a boring prospect, and I mentioned this casually to Breit. A day or so later I got a call from Miles Hanley, the dean of linguistics studies at Wisconsin, who offered to let me take his graduate course as a substitute for the prescribed English course. It was a fortuitous event, as I acquired a lifelong interest in linguistics and the structure of languages (mostly Indo-European, but I have learned a little about the twenty-one Native American languages indigenous to New Mexico). Hanley offered me a graduate assistantship to study the emerging idea of developing a sound spectrograph, but I had no intention of leaving physics. I never told Breit that Hanley had tried to recruit me after the course!

We were to stay in Madison just one semester. When we arrived, Breit called me in and said I should apply to Yale. He had accepted a professorship there, and the group—his graduate students, postdoctoral researchers, and one undergraduate (me!)—was moving.

On to New Haven

The transfer from Madison to New Haven was relatively painless. We found a comfortable room in a co-op house across the street from the Sloane Physics Laboratory. A couple of other physics students lived

there too, as well as students from other parts of the university—many of them from foreign countries. Later we moved into a quonset hut "village" on campus just below the lab. The office of the dean of Yale College was somewhat dismayed to have a transfer undergraduate coming in, and at the second term of the school year to boot. My qualifications were not the problem, apparently. I had taken the standard entrance exam in Milwaukee and found it easy, and my grade point average from the five other universities I had attended was 4.0. It was just that before the war, transfers were rare, and at mid-year, unheard of. I'm sure that with the way students migrate now, even Yale no longer finds such transfers unusual.

I chose courses to balance my program. I have always enjoyed the humanities and social sciences, so this was no burden. In an advanced Shakespeare class, I wrote a term paper on one of the plays in which I proposed an emendation to a line that was not in the Variorum Edition. My teacher liked the idea, but said there was no way to check it—as I had understood in advance! I read Sommerfeld's *Atombau* for German, and for a psychology course in propaganda wrote a paper describing the misdirection we used at Los Alamos to shield our activity. The citizens of Santa Fe were curious about our activities on the Hill and sometimes questioned us, offering free drinks in the bars. To make sure we didn't chance giving away classified information, we invented stock stories. One I liked especially was that we were building special submarines to be launched in the Rio Grande at Española—where the water is about six feet deep in a good spring runoff!

Yale gave me a BS with "Philosophical Orations" in the spring of 1948. At that time the university used both the cum laude honors as well as orations honors, but it has since discontinued the latter. Philosophical Orations recognized that I had accumulated one of the four or five highest grade point averages in the largest class with the largest number of honors graduates to that time in Yale's history. (Several wartime classes were put together in 1948.)

I had been allowed to take all the first-year physics graduate courses in my senior year, so I was prepared to take the doctoral qualifying exam during the summer after graduation and enter graduate school already "qualified." My paper on penetration of gamma rays was published my senior year, as well as one written jointly with Professor Breit and one of the graduate students (Arthur Broyles), so I guess I really was qualified.

I was awarded my PhD in 1951 in a ceremony that was held outside so that all the degrees could be awarded at the same time. It was the first commencement presided over by A. Whitney Griswold, Yale's new president. Six of us were designated to go to the podium and receive packets of diplomas to be handed out individually by the graduate dean in the courtyard of the Hall of Graduate Studies. I was supposed to stand outside the assembled graduate seating and lead the other diploma carriers forward at the right moment. The proceedings were conducted in Latin, which fortunately I understood somewhat, and I was given cue Latin phrases that were to prompt me. Griswold started the statement in Latin, stumbled over the second sentence, and paused. He started again, and again stumbled. I guessed that most of the audience would not know the difference, so I stepped out, followed faithfully by the other carriers. Just as I reached the podium, Griswold started again—so I lined up the carriers as if that had been intended and waited for my cue. It finally came, and we walked up the steps. When I reached Griswold for my handshake and packet of diplomas, I said, "Congratulations, Mr. President. I didn't think you'd make it." Griswold glared at me, but shook my hand anyway. Later, when I was on the faculty, we had a laugh over my presumption.

The Breit Group

By 1951, our group of postdoctoral fellows, graduate students, and young faculty (Bob Gluckstern and I) was comfortably established in Sloane Laboratory and as "The Breit Group," in the Department of Physics at Yale. We didn't consciously separate ourselves from others in the department, and no one seemed to put us apart, but our association with Breit gave us a special aura—sometimes as objects of pity. Gregory Breit was at the height of an extraordinary career that would span nearly fifty years and lead the way in a number of significant fields of twentieth-century physics, and we were certainly privileged to work with him.

Trained as an electrical engineer, Breit did his early work in radio, including defining the characteristics of early tubes and finite coils. The most important of this work, with Merle Tuve, was in using radio to demonstrate the existence of the postulated ionosphere. Bouncing radio waves off this layer of the atmosphere would reveal its existence, and if the time of the return signal could be measured, the height of the layer could be determined. Eventually, Breit and Tuve thought to pulse their signal in order to leave time for the return signal to be recognized, and thus they invented the principle of radar.

Breit was the first American physicist to realize that inducing nuclear reactions with artificial sources—particle accelerators— would be superior to the use of naturally radioactive sources pioneered by Ernest Rutherford. The first nuclear projectiles used to probe nuclear properties were accelerator-produced light ions: the proton (ionized hydrogen), the deuteron (ionized heavy hydrogen), and the alpha particle (ionized helium). An ion is an atom with at least one electron stripped off, so that it has an unbalanced (positive) nuclear charge. (Negative ions, with extra electrons, also occur, but they are not of present interest.) A charged particle may be pushed by an electric or magnetic field to a speed high enough to allow it to penetrate the repulsive barrier of the target nucleus—and thus probe the nuclear properties. (The neutron faces no Coulomb barrier, since it is of neutral charge, but it can't be directly accelerated either.) Two Cantabrigians, Ernest Walton and John Cockcroft, working independently, beat Breit and Tuve by a couple of months in publishing the results of the first accelerator-produced nuclear reactions.

The first nuclear accelerators were linear, but Ernest Lawrence soon invented the cyclotron, where the ions spiraled out from the center of a circular magnet whose field, perpendicular to the path of the ions, kept the ions traversing semicircles. At every half turn the ions got a boost from an electric field. Obviously, any accelerator needs a means of stripping off an electron from the projectile before it enters the accelerating fields. It turns out that the critical parameter for the cyclotron is the ratio of charge to mass of the projectile. For the proton, the ratio is 1/1: one unit of charge divided by one unit of mass. For the deuteron and alpha the ratio is 1/2: one unit of charge to two of mass for the deuteron, and two units of charge to four units of mass for the alpha.

During the war, Lawrence's devotion to driving ions in circular paths had inspired him to propose a machine called the calutron as a means of separating the isotopes of uranium. The calutron, like a cyclotron, has circling ions but has a path in the shape of a racetrack. Ions of the two main isotopes of uranium traveling perpendicular to the magnetic field of the calutron would move in paths of slightly different radius, since they had the same charge but different masses. Thus they could be collected at different points after traversing the magnetic field. This was an old method of separating isotopes, but until the development of the calutron it could not be used to produce macroscopic amounts of the desired isotope. A number of the calutrons that had been used during the war at Oak Ridge were converted into particle accelerators and used in the first experimental explorations of heavy ion reactions.

The term heavy ion refers to ions heavier than the alpha. In 1952, Breit, Bob Gluckstern, and I proposed that there was some interesting physics that could be done with heavy ions as projectiles. (This would include the properties of nitrogen/nitrogen interactions, which we studied in connection with the hydrogen bomb, described below. In fact, it was our study of the nitrogen/nitrogen reaction that gave us the idea of heavy ion nuclear physics.) If all the electrons could be stripped off some of the atomic elements heavier than helium, they would have the charge to mass ratio of 1/2 or 2/4, the same as the deuteron or alpha particle (lithium: 3/6; beryllium: 4/8, boron: 5/10; carbon: 6/12; nitrogen: 7/14; oxygen: 8/16, and so on). This means that they could, in theory, be accelerated in a cyclotron designed for deuterons or alphas, or, if one designed a heavy ion cyclotron specifically, all these species could be used as projectiles!

Since Yale had only a small cyclotron, we decided to ask for funding to build a heavy ion linear accelerator. We had developed a method of following a charged projectile theoretically through a linear accelerator using matrices and offered as part of our proposal a preliminary design. We pointed out that in addition to studying problems in nuclear structure, the heavy ion would be an excellent tool for making transuranic elements. If one could drive, say, an oxygen nucleus into

uranium, one would get a nucleus with eight additional protons, that is, a new element eight places beyond uranium in the periodic table! Of course, neptunium and plutonium are transuranic, but they were made with neutron bombardment, so only a couple of places beyond uranium were achieved. As the elements get heavier, they become more unstable—their half-lives for radioactive decay become shorter and shorter—so a gradual build up by neutron addition becomes impractical.

The University of California at Berkeley piggybacked on our proposal, and we were both awarded funds to build linear accelerators—ours for nuclear structure studies, theirs for transuranic element making. When the grants were awarded, Bob Gluckstern, who visualizes electromagnetic fields in his sleep, drew up much of the final design for both accelerators. Today, heavy ion nuclear structure physics is done with both cyclotrons and linear accelerators across the world. All the two dozen or so transuranic elements made since the war involved heavy ions. Well beyond anything we visualized, today there are relativistic heavy ion colliders exploring energies near that of the Big Bang.

At Yale I also worked on nucleon-nucleon scattering, that is, protons against protons and protons against neutrons. One of my early jobs as a member of the Breit Group was to help prepare tables of functions to describe these reactions, and this required a great deal of computation and some new mathematics. Breit was an excellent mathematician, and I became an adequate one working with him. (It was typical of him that he never actually taught me any mathematics. I absorbed it during our discussions. To him, mathematics was just a tool; we talked about *physics*). In support of this work, we extended the knowledge of the confluent hyper-geometric function, which is the solution of the Schrödinger equation for the interaction of charged particles. *Modern Analysis* by E. T. Whittaker and my old acquaintance, G. N. Watson, provided my guide.

At first we worked with several human "computers"—people working Marchant desk calculators—but soon we were able to extend our computation by borrowing the IBM 602 from the business office and using it as a differential analyzer. My introduction to wiring plug boards for this purpose by Nick Metropolis at Los Alamos came in handy. By the time I was a junior faculty member, we were able to get time on much more powerful, general-purpose computers capable of using stored programs of arbitrary complexity—true von Neumann/Turing machines. These were IBM computers in centers at NYU and at the IBM research laboratories at Poughkeepsie. (Of course, the desktop computer I'm writing on is more powerful than these room-sized behemoths.) The catch was that we were only given access to the machines in the middle of the night. So about once a week my graduate students and I would leave New Haven at 9:00 p.m. and head for the downtown campus of NYU or for the IBM labs in upstate New York. Next morning it was back to New Haven in time for me to teach a class—and for the students to go to bed!

The computers we borrowed at NYU and Poughkeepsie used magnetic tapes for input, and the tapes were prepared from punched cards. Our gradient search in multidimensional phase shift space (as many as forty dimensions!) required us to change tapes at certain points during the computation, and I had to input a parameter at the console. I timed the parameter input so that the program, with changed tapes, was ready for it. We had no programming languages then, but there was an assembly program that made our machine language programs fit the machine. I got very good at octal arithmetic, since I had to estimate the next value of the parameter to try and input it directly as an octal base number. Having been a gun captain for a 155mm coast artillery battery at Mississippi State, I organized my students like a gun crew. At my command, tapes were unmounted and remounted, drives were changed, and so on. I remember looking up at the observation window of the computer room one morning at 2:00 or 3:00 a.m. to see several NYU staff watching us and looking quite amused. Apparently our tape drill had become the object of stories in the computer community.

Our work on the nucleon-nucleon interaction led us to the meson. It had been hypothesized by a Japanese physicist named Hideki Yukawa that all particle interactions were mediated by the exchange of a lighter particle, as electron-electron interactions are mediated by the photon. The short range of the nucleon-nucleon interaction dictated that the exchanged particle be of intermediate mass between the electron and nucleon. Such particles are called mesons, and it could be theorized that in order to mediate the nuclear—or strong—force, they would need to have a mass of about 240 electron masses. The first mesons found in cosmic rays had too small a mass and did not interact strongly with nucleons, as expected, but in 1947, when physicists could return to basic research, the heavier meson was found—also in cosmic rays. The first, lighter meson was labeled mu, and the heavier, pi. (By modern classifications, the mu is a lepton rather than a meson.)

We realized that because, by definition, the pion must interact "strongly" with nuclei, it could serve as a new probe of nuclear structure, which was one of our interests. With the encouragement and assistance of our colleague, the eminent experimentalist Vernon Hughes, we developed a funding proposal for an accelerator we called the meson factory. Vernon was interested in muons (he was the dean of mu meson physics until his recent death at age eighty-three), and the best source of muons is from the decay of pions. Pions decay into a muon and a mu neutrino, so a beam of pions would become in time—as measured by distance along the accelerator path—a beam rich with muons.

In our proposal we created a preliminary design of the factory, with Bob Gluckstern taking the lead, and then had to write a justification for the project—what experiments would we propose to do with the pi's and mu's? I was assigned to write this part, with input from all colleagues involved. I recalled that I had once heard Fermi mention that particles heavier than electrons would be better for treating cancer since they would do less damage to healthy tissue. The gamma ray and electron deposit their energy through a long path, thus injuring tissue on the way to the cancer. A heavier particle, however, would deposit its energy in a narrower region and, with its greater kinetic energy, could be directed deeply into the target. I thus got the idea of proposing the pi meson as an instrument for treating cancer. We could not work directly on medical therapies, but we thought a broader program of research would be helpful in obtaining support. Modern proton beams extend this idea even more effectively.

When we were invited to present our proposal in a conference with funding agencies, Louis Rosen rose after the Yale group had presented and said, "I want a machine for Los Alamos like Yale has proposed!" As it turned out, Yale didn't get the accelerator, but Los Alamos did; and they gave Vernon first use of the muon beam for several years (without the administrative grief of running the show). Once again, Bob Gluckstern worked on the electromagnetic fields for the final design. When I returned to New Mexico years later, Louis showed me the accelerator. As we looked over the experimental area from an overhead catwalk, I could see the setups for some of the experiments I had written about in the scientific justification.

A colleague in Yale's medical school, Morton Kligerman, decided to try pi meson cancer therapy and started The Cancer Research and Treatment Center at the University of New Mexico Medical School, only ninety miles from the accelerator at Los Alamos. I understand that the pions worked as advertised, but the treatments were discontinued as too expensive compared with other treatments available. The CRTC continues as a part of the UNM medical school, but Kligerman left the Center when the pion experiment was canceled. Years later, when I returned to New Mexico, I became a patient at the Center for continued monitoring of a colorectal cancer that had been removed in Buffalo. (I was discharged last year; by now there's little chance of that cancer returning.)

Gregory Breit

My association with Gregory Breit would last twenty years with day-to-day interactions and to the end of his life as frequently as possible thereafter (I was actually chair of his department at SUNY Buffalo), so I can attest as well as anyone both to his genius and his irascibility. Breit had a dark side that was the source of legends even from his New York University days. He was notoriously prone to outbursts of temper, though he routinely apologized for them—quite sincerely, I believe. I experienced only one of these, very early in our working relationship, when I was still an undergraduate. Breit had lost his temper over a misunderstanding about the way I had edited a joint paper. When he blew up, I said something like, "I think I'll return when you are feeling better. Please call me." I walked out and went back to my office. He duly called with his apology, which became even more heartfelt later as I explained the misunderstanding. He never raised his voice to me again. We had our differences over the years, but they were always resolved without temper—albeit with some tension at times. We became very close friends over the succeeding years. (Bob Gluckstern's wife made the observation that Bob and I could tolerate Breit's behavior because we both had difficult fathers.)

Others of my colleagues were not so lucky. I was present when Breit took the hide off a graduate student who had wished him "a good talk" at a meeting: Of course his talk would be good! And it wasn't just his students who were regular targets of his temper; colleagues came under fire as well. I heard about a heated "discussion" between Breit and Henry Margenau on the occasion of a seminar in honor of Breit's retirement from Yale, when each thought the other's talk was the "worst ever heard." Having heard the talks, I fear both were correct!

Breit's solicitousness was equally legendary. He was formally polite in the European way as a result of his upbringing in the first decades of the twentieth century. He always stood when you entered his office and expected you to do the same if he visited yours. When I decided to go to Wisconsin at his invitation, one piece of advice I was given was never to get behind him and Eugene Wigner as they approached a door. Both were trained in European politesse, and it was nearly impossible for them to get through, as each would defer to the other until someone else broke the impasse in desperation.

Breit was known for his devotion to his students' intellectual development and personal welfare. He was available at any time for consultation, and if a student were shy, he would be invited in for a chat. At our weekly group lunches we were terrified that Breit would ask one of us a difficult question, but we learned a great deal, including how to think on our feet. Frequent "parties" at his home, set up by his wife, Marjory (who also entertained the wives of group members, separately), were wonderful opportunities to talk physics in general. Breit was incredibly well informed; he received hundreds of preprints a month, read most of them, and shared them with the group members according to current interest. He appeared to retain everything he read, including the location of the source—journal reference, page number, even location on the page!

The sessions at Breit's home also were opportunities for us to meet the great physicists of the time. He invited many notable visitors as they were available. I especially remember meeting Werner Heisenberg, for example. Someone had told him I had been at Los Alamos during the war, so the head of the "counter project," as we at Los Alamos called the German bomb effort during the war, asked me a number of questions— most of which I couldn't answer for security reasons!

The phrase idée fixe could characterize some of Breit's responses outside of physics. Once, early in our relationship, he asked me to work on a calculation on a Saturday afternoon. I explained that I couldn't do the job just then, but would do it Sunday. Why the delay? Yale plays football Saturday afternoon, and I usually attend, I told him. For the rest of our association, he assumed I went to football games on Saturday afternoon—spring, summer, fall, winter, year after year! I never corrected his impression, for it made my life easier at times.

The Super

Sometime in 1951, Breit called his group into his office to discuss a request he had received that we take a look at the current work on the hydrogen bomb, or "super," as I still called it from Los Alamos habit. There had been much discussion of the super toward the end of my stay in Los Alamos. During the summer of 1946, even we youngsters had had many bull sessions about the possibility both of building a hydrogen bomb and eventually controlling fusion for the purposes of generating energy. We were confident that the super would be working in perhaps five years. (We believed that peacetime energy production would follow within ten years. As I write, almost sixty years later, we still haven't got fusion energy controlled for power.)

Rutherford, using the accelerator designed by Cockcroft and Walton, had used deuterons (ionized deuterium atoms) to bombard a deuterium target (D+D) and, among other reactions, found that sometimes the two deuterons fused into helium with the release of considerable energy. This was in 1934, before anyone had thought of nuclear weapons, and so the result was treated simply as an expected reaction. Similar experiments showed that tritium (hydrogen mass 3) and helium 3 could also be made, and thus Rutherford found all three isotopes he had postulated. The mass 3 isotopes would turn out to be very important for fusion weapons.

The fusion of two deuterium nuclei releases an enormous amount of energy. Recall that we are dealing with the binding energies of nuclei, and that iron, with an atomic number of 26, has the highest binding energy per nucleon, the total binding energy divided by the number of nucleons. A fission explosion releases some of the binding energy of a heavy nucleus by breaking more weakly bound heavy nuclei into fragments of more tightly bound elements, which reside in the middle of the periodic table. In suitable units the release is about one unit per nucleon of the fissile element, or about two hundred units per fission. But if there is a maximum at iron, it can be approached from either direction. Thus there would be binding energy released if lighter elements were fused into heavier elements with greater binding.

A notable case is helium, the end result of the fusion of four hydrogen nuclei. Helium contains only two protons and two neutrons, but its binding energy is off the curve at about twenty-eight units, or seven units per nucleon. It is more tightly bound than a number of heavier elements. (The unusual binding of helium is partly the result of the action of a symmetry law that Wolfgang Pauli had found in atoms.) In the late thirties, Hans Bethe had shown that a series of reactions among the light nuclei, beginning with D+D, could lead to the equivalent of the fusion of four protons into helium. (In this prototypical elementary reaction, two of the protons would absorb electrons and become neutrons, also producing a couple of neutrinos.) The initial reactions could be driven by the gravitational energy of the assembling hydrogen of the star and then by the thermal energy produced by the reactions themselves at solar density. Thus the astronomical heat and light of a star could be explained. If we could assemble the helium nucleus from its free constituents, we would gain a factor of seven in efficiency over fission.

The actual formation of helium from four protons is unlikely even in a star, however, which is why Bethe hypothesized a cycle of several intermediate reactions that, cumulatively, have the same effect. It is more likely that deuterium was formed first after the Big Bang. A regular hydrogen nucleus contains a solitary proton, but the hydrogen isotope deuterium also contains a neutron. Thus in a deuterium reaction just a single encounter is required to produce the two protons and two neutrons that make up the helium nucleus, as Rutherford had shown. This is one way we think helium was formed after the Big Bang.

The Bethe cycle could not be used for a fusion weapon because some of its reactions are too slow. But there were other reactions that could work, especially if the explosive contained some nuclei further along in the cycle. Thus there are better candidates than deuterium for fusion reactions on earth. Another, heavier isotope of hydrogen, tritium, has one proton and two neutrons. Helium-3 also exists, as Rutherford had shown, with two protons and one neutron. Deuterium plus tritium or deuterium plus helium-3 are both better than deuterium plus deuterium to produce helium-4. Less binding energy is released than for the theoretical four-proton reaction, but the efficiency is still several times that of fission.

These mass 3 isotopes are not readily available in nature and so must be manufactured. Helium-3 can be separated from normal helium, and tritium can be made in a nuclear reactor. While bombardment of deuterium with reactor neutrons is feasible, an easier reaction occurs if we bombard lithium-6. It absorbs the neutron and then breaks up into tritium, ordinary helium, and some energy. The last neutron is not tightly bound, so tritium is radioactive. With a half-life of over twelve years, it is stable enough to be used, but obviously presents a problem for the maintenance of stockpiles of weapons. As usual, nature gives us problems: the mass 6 isotope of lithium is the rare one, just as is the fissile isotope of uranium.

The principal technical problem in building a hydrogen bomb is how to achieve the temperature and pressure required to initiate a self-sustaining fusion cycle, since the repulsive Coulomb force between positively charged nuclei must be overcome by the kinetic energy (equivalent to temperature) of the reactants. Particle accelerators can, of course, produce any reaction of the cycle—but only one at a time. Hence they are useless for acceleration of a macroscopic sample of fuel to the energies (or temperature) necessary to initiate fusion. Baratol or Comp B can compress plutonium sufficiently to produce the supercriticality necessary for a high order nuclear fission explosion, but they can produce nowhere near the compression needed for fusion.

The fission bomb itself produces a very high temperature and a significant pressure wave, as we have seen, so it could act as the "fuse" for a fusion assembly. But the use of a fission bomb was not quite as trivial as initially thought. Calculations showed that it would not work as first conceived by Edward Teller. Stan Ulam, a mathematician who had been at Los Alamos during the war, thought of using the neutron flux from the fission bomb to exert pressure on a special hydrogen mixture (deuterium/tritium, in liquid form during the early studies). His calculations showed that, suitably set up, this pressure plus the heat from a fission bomb should ignite the hydrogen. Teller, responding to Ulam's idea, thought that the radiation pressure could be even more effective. (For modern fission bombs, this is especially so, as much more of their energy is in radiation than was the case with the early devices). The resulting design is usually called the Teller-Ulam weapon. One of the problems Teller asked us to work on was to confirm that the radiation pressure of a fission explosion would be sufficient to ignite a hydrogen bomb. My work on penetration of gamma rays fit right into this problem, so I did the calculation, verifying the positive result I assumed had been obtained by much more senior physicists at Los Alamos.

This is how a hydrogen bomb works: An implosion fission bomb produces the temperatures necessary for a deuterium/tritium mix to interact, initiating the fusion explosion; and the pressure of the bomb's radiation compresses the mix sufficiently for the reactions to take place before the whole device blows apart. If the reactions can be made to go, about twenty-two units of fusion nuclear energy are released per helium nucleus made. This may not be quite as potent a process as the assembly of free nucleons, but it is much more efficient than fission. To a weapons designer, there is also the advantage that no chain reaction is involved. Fusion is the same as burning: if you wad the piece of paper up so the heat can be transmitted easily from one point to the next in the rest of the ball, you only need to apply the match (at the temperature of Fahrenheit 451, made famous by Ray Bradbury and revived, *mutatis mutandis*, by Michael Moore) to one point to start. Thus, theoretically, you can make a hydrogen bomb as large as you wish, since there is no limit to how much fuel is used. Still, practical considerations involving the spontaneous disassembly of the fuel, and perhaps the sheer size of the device, limit the fusion bomb to fifty or so megatons TNT equivalent—some two thousand times the power of the first fission bomb.

Cold War Security

By 1951, we were in the thick of the Cold War, and all work on the super was classified, calling for clearances. My own Q was still in force, giving me access to any document I needed; and of course Breit, as the first head of what became the Manhattan Project, was still cleared. But we needed a secure place to work. A partition was thrown across one end of a large room in Sloane, and thus the Secret Room was created. It was obviously not secret from anyone in physics, but what went on inside gave it the name. Security was supplied by ADT, and only Breit, Gluckstern, or I could open the door at the beginning of the workday, after the requisite telephone call.

There was a vent near the ceiling at one end of the Secret Room, through which outsiders might overhear our conversations. I was given the task of fixing this security breach. I was not the carpenter then that I have become since, so the solution of boarding up the hole never occurred to me—or perhaps we needed the vent for air circulation. In any case, I decided on a "high tech" solution: an audio oscillator to be placed in the vent and turned on whenever we were discussing the work. I was aware that the Weber-Fechner law implied that a constant tone could eventually be heard through, so I thought we needed an oscillator with a variable pitch. (Nowadays this is how alarm signals work, but not in 1951.) As a theoretician, I know how things work only in a most direct way, so I designed (and built) an oscillator with a partial disc rotating through a conducting cavity such that the resonant frequency changed periodically. To get sound in the middle of the audible range, I made the variable capacitor about the size of Lawrence's first cyclotron. The device worked—and gave off the most horrible sound it has ever been my misfortune to hear! We used it as infrequently as possible. The fact that my friends outside the group didn't cut me off from their society shows their generosity of spirit.

The file cabinets needed secret combinations, and Bob Gluckstern suggested natural constants. I told him of how I had once startled a security officer at Los Alamos by confidently offering to open an absent colleague's cabinet to which he needed access. I tried pi, then e, then gamma, and so on. One finally worked, and the next day we got orders to change our combinations. I changed mine to the seventh through twelfth digits of pi. Gluckstern thought that was a good idea. As a former math club contestant in New York, he knew several natural constants to a dozen decimal places and rattled off the digits. I, of course, would have to look them up in a table until I memorized them. My first try the next day brought no result. We finally determined that Gluckstern had slipped a digit. I could still look up the combination if I forgot, but I had to put in a six instead of a five.

Gregory Breit, I, and Bob Gluckstern, in that order, were to be contacted by ADT if there was ever a security breach. One morning about 3:00 a.m., I got a call. Come to the lab immediately! When I arrived, three ADT men were outside the door of the Secret Room with drawn revolvers casually aimed at John McHale, who was sitting somewhat uncomfortably in a chair in the midst of the guns. John had been unable to sleep and went to the lab to do some work. He forgot that the door to the room had to be opened the first time each day by an authorized person in an authorized procedure. He had never been around when Gluckstern or I opened it. I cleared things up, thanked the ADT men for their alert attention, and went back home to bed. Fortunately, that was the only "breach" we experienced as long as the Secret Room existed.

Could We Endanger the World?

The central question Edward Teller had asked Breit to consider was whether the hydrogen bomb would ignite the atmosphere. Bethe had done some work on this question before the Trinity Test with respect to the fission bomb, but the super was expected to be a thousand times as powerful and to involve reactions closer to the constituents of the environment. By 1951, others had already done some work on atmospheric ignition for the super as well, but Teller wanted the "most careful physicist he knew" to do the definitive calculations. So Breit got the call.

As scientists, we appreciated that it was simply a matter of normal scientific curiosity that the Soviets would attempt to develop nuclear weapons. It would also be a matter of national pride, and perhaps even of national survival in the presence of a nuclear-armed and hostile U.S. By the same token, any one of us who understood the scientific basis for the work of the Manhattan Project expected the Soviet scientists—perhaps spurred on by the necessity of satisfying a vicious and deadly leader—to develop a nuclear weapon fairly quickly; for the most important "secret" was that one could be built. I didn't know of the assistance that my former colleague Klaus Fuchs had given the Soviet effort, but I assumed that nuclear weapons in the hands of a hostile Soviet government were simply a matter of time.

Still, it was difficult to imagine what would be a suitable target for a weapon as powerful as the super—New York, Chicago, Los Angeles, Moscow, St. Petersburg . . . ? Work on the fission bomb had continued at Los Alamos since the war, and we were already building a stockpile (sometimes haltingly, as I was to learn much later) of nuclear weapons capable of destroying the world in a very short time. The super would reduce that time to just hours, perhaps, if it ignited the atmosphere. Why, then, try to make such an apocalyptic device?

Many scientists—Oppenheimer most prominent among them— refused to work on the super. I was aware that Oppenheimer had persuaded the advisory committee he chaired to recommend against a crash program of research and development on the hydrogen bomb. While I admired his principles—as I admired him—I doubted the program would be stopped.

I would have liked to join Oppie in opposition to the super, but only if we could have been sure it was not technically feasible. To my mind, the sad fact was that such a weapon was too attractive to single-minded militarists and aggressive governments to be ignored, and thus it would be made somewhere if scientifically possible. I believed that the U.S. was the best country to develop the super first, because it would be least likely to misuse it, and our possession of it would deter more irresponsible nations. This was assuming, of course, that the motive of the U.S. in building a super was defensive and not aggressive, which would be consistent with our foreign policy since the founding of the country.

Edward Teller had been campaigning for the development of the super for some time and was now pushing for a major effort—with him in the lead. His preoccupation with the super at Los Alamos during the war had kept him from participating significantly in the development of the fission weapon. This was despite efforts by Oppie to get him involved, we heard. Teller's unwillingness to work on what the rest of us were breaking our backs over made him unpopular among the younger people I knew. I had no personal data to judge the general opinion that he was jealous of Fermi, Oppenheimer, and Ulam, but that would explain (along with Oppie's opposition to development of the super) his disgraceful behavior prior to and in the Gray committee hearings on Oppenheimer. He helped a petty man, Lewis Strauss, to harass a man better than either of them. Of course, General Groves was as culpable; he had defended Oppenheimer as long as he was useful but abandoned him when his (Grove's) military career seemed threatened. Teller's behavior in trying to keep any credit for the design of the super from Stan Ulam was equally despicable. When, as dean of Graduate and Professional Education at SUNY Buffalo, I had to introduce Teller years later, I characterized him as honestly as I could without being rude. I said only that he was one of the best teachers I had ever encountered (at Los Alamos "University"), a true statement, which Teller liked.

I was willing to work on the problems Teller had presented to Breit, but I would do it for the country, not for Teller. My point of view about the scientific development of weapons was arrived at independently of any of my Los Alamos seniors—or of Breit, for that matter. (Breit was a staunch anticommunist, as were many other Eastern European refugees.) My view was based on an analysis of the scientific and the political situations as I understood them and a faith in the integrity of this country and its adherence to the enlightened humanism that imbued its founders. Of course, idealism is sometimes hard to maintain in face of reality!

I wanted to believe that if theoretical studies of the super showed it could indeed ignite the atmosphere, then no sane person or government would proceed with an effort to make one; but I realized that this was no guarantee that it wouldn't be made. We had just ended a war that some insane people had started, and I was well aware that there was (and is) a fairly widespread willingness among political leaders to ignore scientific advice that is inconsistent with their personal agendas. What is more, I had long thought of Soviet Communism as a secular religion, and I knew from history the depredations that had been committed in the name of one religion or other. It no longer seemed unthinkable to me that the horrors of Hiroshima and Nagasaki could be repeated—on an even greater scale.

I don't know what my colleagues in the Breit group thought about the super, but I'm sure it was clear to all of us that the question of whether it would set off the atmosphere if tried was a matter of species survival: if the atmosphere was ignited, all life on earth would be eliminated! I was the only one of the group who had worked on the fission bomb, so the others probably had not really thought about the personal responsibility attendant to working on apocalyptic weapons. But in any case, we went to work with a will.

The question of igniting the atmosphere with a fusion device involves the same nuclear reactions, with nitrogen and oxygen, as one would consider in the case of the fission bomb, but the temperatures are higher, the material velocities greater, and the radiation more intense. Nitrogen is the largest component of the atmosphere—78 percent—so nitrogen-nitrogen reactions are the most likely to sustain ignition. Oxygen, at 21 percent, would be important also if the atmosphere actually began to burn, but nitrogen is likely to ignite first because its nuclear charge is seven, whereas oxygen's is eight. The lower the nuclear charge the lower the Coulomb barrier, the repulsive force between like-charged particles. Just as with the fusion reaction itself, this repulsive force must be overcome by the energy (and hence velocity) of the interacting particles if a reaction between atoms in the atmosphere is to take place. Thus we studied only the nitrogen/nitrogen interactions to begin with.

Two nitrogen atoms flung at each other with enough energy will produce one of several possible outcomes, depending on the details, and on balance will consume more than half the energy of the fusion reactions of the bomb itself. So a primary question is whether the energy available from the most energetic nitrogen/nitrogen reactions is enough to sustain the burning; if the nitrogen-nitrogen reaction has too little exothermic energy, the reaction will fizzle. Of course the density of the atmosphere is decidedly less than that in a bomb, which also works against equilibrium burning. Still, we could be confident that the super would not ignite the atmosphere only if the magnitude of the fusion explosion were such that the environment it produced would stay well below the threshold for any nitrogen reaction, and this turned out to be the case with the parameters we were given. We determined that the temperature of the explosion produced by the super would be much too low to initiate nitrogen burning.

We worked on our calculations for several months, with all results communicated to Breit without copies. He wrote whatever final reports there were. Our recommendation was secure only if all the cross sections were correct. (The natural reaction rates for all expected reactions are typically expressed in units of area of the order of magnitude of the physical dimensions of the target nuclei—that is, of the cross section area of the target particles. The unit of cross section for nuclear reactions is whimsically called the "barn.") Thus Breit made the remeasurement of critical cross sections a condition of our conclusion, and I believe they were quickly carried out. Our assessment was borne out when a test device was safely exploded in the Pacific on November 1, 1953, with a yield of about ten megatons TNT equivalent, nearly a thousand times more powerful than the fission bombs of WWII.

Administration

The Breit group gave me excellent training, and eventually I became a regular faculty member at Yale. My situation was almost ideal. I was tenured in one of the best universities in the country—or the world, for that matter. I was doing research I enjoyed immensely with colleagues who were friends. My teaching, both undergraduate and graduate (I had taken over the beginning graduate theory course from Leigh Page, legendary Yale physicist, when he died), was enjoyable, and I was beginning to participate in the broader life of Yale through a fellowship in Timothy Dwight College.

When I was asked to take over the directorship of graduate studies for the department (in those days, essentially assistant chair), I accepted, thinking I might bring some order into the selecting and support of graduate students for physics. We received applications from perhaps three times as many students as we could accommodate each year, and I believed it was important to accept the ablest in as fair a procedure as could be devised. My policy, as supported by the department, was to admit only students we thought could complete our tough PhD program. (Master's degrees were not a priority.) We knew that some of the departments in our sister schools admitted twice as many students as they expected to finish and let academic attrition find the ablest. Competition was cutthroat in those departments as a result. But I wanted our students to help each other learn physics. When I was a student, we organized seminars on new topics not yet in the curriculum (the latest developments in quantum electrodynamics, new ideas in the application of group theory, and so on) and taught each other.

Thus I began my fall from grace and became an administrator—if only part-time at first. It was actually my second go at administration, as in some sense, leading the powder men at S-Site was administration. I had had to get men to work for me as I directed and deal with people higher up to get supplies, tools, and amenities for them, which is essentially what any administrator does. Like all organizations, a university is a system comprised of interacting subsystems. As an administrator at any level, you must understand the whole structure to get the resources you need to carry out your responsibilities—to the organization itself and to the people in it who are within your scope.

When the first request to take an administrative job outside Yale came (as a department chair at Oregon State), I had to balance my love of the life of a faculty member—by far the best job in a university—against a sense that I might be able to multiply whatever talents I possessed by developing programs and supporting faculty and staff in a university department. I was aware that a university administrator, to be successful, needs to have grown up in the singular culture of this nine-hundred–year-old institution, the descendent of Plato's academy, the repository of the knowledge and sometimes the wisdom of humanity, the engine of discovery about the world and its inhabitants. It helps to have some sense of the unity of knowledge in the midst of the centrifugal forces of modern specialization. I thought it would be worthwhile to take on administrative tasks if I could have the opportunity to help develop a single example of this important institution.

I firmly believed (and still do) that no one who *wants* to be a chair or dean or provost in a university should be appointed, for such a person is usually seeking power. An effective administrator must understand the academic and social environment of the institution and lead by promoting and managing the self-interest of others, putting his or her own self-interest aside. It is the job of the administrator to see that Adam Smith's economic theory is followed as closely as possible in the university setting—self-interest in the common interest. Arbitrary power does not exist in a university—or not for long if tried. An administrator must understand the aspirations of all the constituents of a university to engage their help constructively. He or she must become familiar with the political dynamics that inevitably arise within any human organization to make sure internal university politics don't negatively influence university development. (Woodrow Wilson once said that Princeton University politics were worse than Washington's.) How well or badly I managed to do this is best learned from my colleagues.

I found the complexity of administration to be an interesting challenge, particularly when I reached the provost level (at the University of New Mexico), where the whole academic program of the university is one's (shared) responsibility. Whatever creativity I possess was engaged every moment. No two days were the same. The analytical rigor needed for physics, coupled with the need sometimes to visualize an outcome intuitively, is good training for a university administrator. The modus operandi I developed was straightforward: first seek all available information and all advice that can be acquired in a reasonable time, and then, when reason has done its best, take the remaining intuitive leap. One will never have enough information to *guarantee* the best decision. One must always risk error, because not making a decision is in effect a decision—with no control of the outcome. It takes even steadier nerves to be provost, I found, than to cast high explosives in odd shapes!

Postscript

Gregory Breit was retired by Yale when he reached sixty-eight, and he went to SUNY Buffalo, where I served for a while as chair of physics and then dean of Graduate and Professional Education. He eventually retired to Oregon, where he died in 1981.

Epilogue

The work of the Manhattan Engineering District has had a number of consequences for the world beyond ending the Second World War. Weapons development and production have continued apace at the sites opened in WWII plus others set up to assist them. Hydrogen weapons have entered the production stream, with new ancillary support programs—to produce tritium, for example—in place. Today the work of the weapons laboratories is done by scientists and engineers who have chosen to make their careers in weapons—a possibility opened up by the realization of governments, beginning with ours, that the close collaboration of science, development, and production, modeled on the wartime effort at the labs, is the only way to maintain weapons at the boundaries of what is technologically possible. A number of my former students are among those working at one or another of the labs.

Some things don't change, of course. Any collection of scientists will insist on doing some basic research, and the labs have encouraged that. Basic nuclear physics has been extended, of course, as has as molecular biology and areas of mathematics such as number theory. And applications other than weapons are being pursued. Environmental models to investigate the "nuclear winter" hypothesis of Carl Sagan and associates have been developed at Los Alamos. They have strengthened the evidence for a severe global temperature drop from fires started by a maximal nuclear war. Some of the most advanced work on alternative energy sources (alternative to fossil fuels, that is) has been done at Sandia National Laboratories in Albuquerque, as well as at Los Alamos. New types of particle accelerators have been developed and put to use in areas outside of physics, such as cancer treatment. Even the geology of the Rio Grande Gorge, which was deeper than the Grand Canyon before it was filled in by volcanoes, has been explored. These are only a few examples of the scientific activity that goes on today at the weapons laboratories.

Government-funded scientific activity at universities has also expanded since the war. The urgency of winning World War II engendered extraordinary innovation by scientists developing new devices and improving older ones in programs all over the country, sponsored by the government under the general authority of the Office of Scientific Research and Development. As a consequence of the success of science and development in weapons, management, and logistics—spectacularly represented by nuclear weapons, the proximity fuse, and radar—governments began to fund basic research at universities. This legacy of science in war has made the U.S. the leading scientific country in the world. Today, government support for university research has been extended beyond physics to the social sciences, medicine, and the arts. Much of my education after the war was supported by grants, as was my research as a graduate student and after. It is a concern that fewer students prepare themselves today to take advantage of these opportunities.

What has not kept apace, however, is the public's understanding of scientific issues. One of the most obvious—and most feared—legacies of the Manhattan District is the nuclear (fission) reactor, now used to generate energy in addition to producing plutonium for bombs. Just as it was the radiation it emits that made the nuclear bomb something much more fearsome than simply a bigger bang—as I came to appreciate from my work on the Bikini Tests—it is the problem of radioactivity that haunts the public image of nuclear reactors. While I share the concern of any reasonable person for the dangers posed by nuclear reactors used for power, I believe the adamant opposition to any nuclear activity by some of our citizens is an emotional response rather than a reasoned one.

As I write, we are endangering the world with our consumption of fossil fuels. Global warming will one day make the planet uninhabitable if we don't stop it; it is the most dangerous threat we face now that all-out nuclear war seems less likely. As for radioactivity, an ordinary fossil fuel power plant emits more radioactive isotopes into the environment than does a well-run nuclear plant. Nuclear waste is more dangerous than coal ash, of course, and needs to be handled carefully, but we know how to handle it safely—and how to store it so it does not reenter the environment. The Waste Isolation Pilot Plant near Carlsbad, New Mexico, recently began to receive nuclear waste without incident. Its presence in my home state is not a worry for me, and neither should it be for the residents of Carlsbad.

It is perhaps the fear of accidents at reactor sites that has increased public resistance to nuclear energy production more than anything. But the likelihood of accidents can be reduced, and the Three Mile Island case showed that even when it does occur, a major accident need not harm the public. Nuclear reactors in the U.S. are surrounded with containment vessels. Such a vessel kept the radioactivity confined during the Three Mile Island accident. Many more patients have been killed by dosage errors in hospitals around the world than have been killed in accidents at nuclear power plants in the U.S.

The problem with the commercial reactor accidents is the same as in hospitals—human error, which, perhaps abetted by lack of deep understanding of the particular danger, has not been eliminated. But it is possible to train reactor operators to such a level of understanding that their actions will not trigger accidents, and they must be paid commensurate with their training and responsibility. We do not pay an airline pilot for the minimal work he or she does during a flight with an automatic pilot, but for his or her training and responsibility. Unfortunately, human error and willful ignorance are natural characteristics we all share. Witness a half-million highway deaths a year or the self-destruction caused by smoking. It is not possible to eliminate human error—or stupidity—from life, but we can develop fail-safe processes in dangerous circumstances.

It is instructive to look at the difference between the Three Mile Island incident and the terrible accident that occurred at Chernobyl in the Soviet Union in 1986. The reactor at TMI had a containment vessel around it; the one at Chernobyl did not. Thus the TMI reactor was, in this case, foolproof (since fools were certainly operating it at the time of the accident), while the Chernobyl reactor was not. The Chernobyl accident provided a perfect example of why many people fear reactors. Not only were persons killed in the plant, but radioactive fallout poisoned the immediate countryside and posed some danger to people in downwind countries. Fallout is a danger that an ordinary power plant explosion would not pose.

It is important to remember, however, that even the accident at Chernobyl did not cause a nuclear explosion. The basic design of nuclear reactors the world over inhibits nuclear explosion. Getting a nuclear explosion to occur is not easy, as I have tried to show, and the conditions in a reactor are far from those in a bomb.

This does not mean that nuclear energy is easy to produce safely or that the problems of reprocessing or storing spent fuel are trivial. But with intelligent design and strict regulation these problems become technically manageable. If the public can be convinced that its problems have indeed been properly managed, then nuclear energy might live up to its potential as environmentally better and potentially more sustainable (with breeder reactors to produce fuel as they develop energy) than fossil fuel. Thus I believe an irrational fear should not keep us from considering nuclear energy on its merits.

The future holds an even better source of nuclear energy in the form of fusion energy. But the scientific problems with fusion are immense, and only a few laboratories have managed to produce controlled-fusion energy. Sandia Laboratory, for example, has made it only in very small amounts and at far below the break-even point compared with energy input. At Los Alamos during the war, we young-sters thought that fusion energy was about a decade away. Almost sixty years later, it is still at least a decade away!

The possibility of nuclear holocaust is the most ominous legacy of the Manhattan District. Sagan and his colleagues have forecast nuclear winter, as we have noted, and Jonathan Schell (see bibliogra-phy) has forecast the elimination of most of the life on earth, includ-ing us specifically, from radioactive fallout from such a conflict. (My own calculations aren't quite so pessimistic, but the few remaining humans, scattered over the earth, would survive at a level of existence reminiscent of the Stone Age.) Neither of these apocalyptic conse-quences is guaranteed, of course; earth's environment is too large and complicated for the calculations to be certain. But they are close enough for sensible people to say that nuclear war is unthinkable.

Unfortunately, world leaders are not all sensible people, and there is an irrational attraction for "joining the nuclear club." The principal dif-ficulty, as it was for us during the war, is getting enough nuclear explo-sive for a bomb. This, therefore, is a point at which global control can be exerted. Plutonium is made in reactors and processed chemically. These activities are difficult to hide, for they cannot be performed in tiny labo-ratories. Separation of uranium explosive from uranium ore is a similar-ly massive operation. Constant monitoring by the world community not only is feasible, but imperative, and negotiations must be started early enough to prevent disaster. Both plutonium and refined uranium are nuclear fuels as well as explosives, which complicates the task, but sound international controls — exempt from politics — can do the job.

Absolute nonproliferation is the answer, of course, but it has been demonstrated to be impossible. There is no real chance of getting the nuclear powers, including the U.S., to get rid of their dangerous stock-piles. But informed citizens — scientists foremost among them — must urge their democratic governments to reduce nuclear weapon stockpiles below the level where species existence is threatened by a total war.

During the Cold War it was believed that the presence of nuclear weapons acted as a deterrent to war in general. Indeed, it is encouraging that they have not been used since World War II. Richard Rhodes (History Channel: *Modern Marvels/Manhattan Project*) has pointed out that the casualties of war have proportionately dropped in the postwar period. But we must not be lulled into thinking that the possibility of nuclear war is remote. Democracy in Russia is not yet ingrained, and nuclear stirrings have been revealed among several so-called rogue states. Any nuclear power might use nuclear weapons if attacked. The United Nations and other coalitions of rational countries must work tirelessly to keep situations from developing where the use of nuclear weapons might seem justified. Preemptive war without a very clearly defined threat was once contrary to American policy, but is no longer so. We cannot allow our leaders even to think a preemptive nuclear strike is possible.

I am uncomfortable with the development of specialized nuclear weapons, such as lower yield "bunker busters" and artillery shells, because they seem prosaic and thus less objectionable to use. There is no acceptable degree of nuclear war, for once launched, no one can control the course of a nuclear war. Nuclear warfare must never be regarded as an extension of conventional warfare, but as a cataclysm to be avoided absolutely by every means at the world's disposal. I am optimistic that nuclear weapons, as a technical, moral, and political issue, can be managed rationally and creatively by world leaders—but only if we, as world citizens, continually demand it.

Since I left Los Alamos in 1946, I have worked in various ways to inform our citizens of the dangers to all of a nuclear war. Early in my academic career I began to teach undergraduate courses in which I included discussions of nuclear weapons and their multiple destructive consequences. One of the most effective teaching tools I have found are films the Army Signal Corps made of Hiroshima and Nagasaki just after the end of the war. I had given a lot of thought to the consequences to people and cities of the use of nuclear weapons, but my imagination needed the assistance of these films to understand the actual range of destruction. This is where I saw the vaporized image of the "running man" on the bridge, which is imprinted in my mind's eye. That particular film is called "A Tale of Two Cities"; I guess it was the best of times because the war was ended and the worst of times because two cities had been destroyed with just two bombs. My purpose in showing these graphic films is to inform students so that they can effectively fulfill their democratic responsibility of monitoring their elected officials and the policies they advocate. This book is intended to take this information beyond the classrooms where I have always worked. No nuclear weapon has been used against any population since Nagasaki, and it must remain that way.

So how, nearly sixty years along, do I view my own participation in the beginning of this threat to life on earth? The history of science has shown that whatever is not forbidden by the laws of nature can—and eventually will—be done. Some of my colleagues have refused to work on weapons, and I support their determination while I do not share it. I believe the United States is more likely to have "sensible" leaders than all but a few other nations in the world, despite some historical and current counterexamples. The counterexamples warn us that constant vigilance on our part and that of our representatives is essential to make sure that our leaders behave responsibly and to remove them if they do not—the essence of democracy. I thus remain convinced the United States must lead in the design of new weapons so that nations who may have a dangerous capability will be restrained. As difficult as it is to contemplate their awesome destructive power, we must manage these weapons rationally. Even had I known the consequences of the bombings of Hiroshima and Nagasaki beforehand, I would still have worked as hard as I did to make the weapons a success. The personal consequence is that I have a share of responsibility for the destruction of two cities and thousands of civilians living in them. That is a responsibility I shall carry with me for the rest of my life.

Coda

Most of my research and teaching has not related directly to the Los Alamos work. My research has focused on the physics of the nucleus, principally the phenomenological study of the nucleon-nucleon interaction. I've taught, with pleasure, the gamut of graduate courses in theoretical physics and those for undergraduates as well. Throughout my university career I have been especially active in developing and teaching courses for nontechnical majors, discussing the intellectual relations among the arts, social sciences, humanities, and sciences as we try to understand the universe and ourselves.

Our children, John and Wendy, were born in New Haven Hospital and returned to New Haven for college. John graduated (after an interruption for service in the Marines during the Viet Nam War) in English and went to Illinois for graduate work in fine art (painting and drawing). He married a fellow art student at Yale, Shelley Gaffin. He is now professor of painting and drawing at Colorado/Denver. Wendy did a double intensive major in Greek and Latin and did her graduate work in Bristol, having married an English vicar, Terry McCabe. She operates a business in needlework designs in St. Neots, Cambridgeshire. Our oldest grandchild, Damaris, works at the Royal Opera House, using her education in arts management while working on a graduate degree in film. Ursula (our second grandchild) is starting A-levels and preparing to play Eliza in *My Fair Lady* this year. Isaac, John's son, plays football for his high school in a Denver suburb. So I think we have had a reasonably normal life alongside of physics and university administration (though whether the others would characterize it that way is another matter).

Los Alamos frequently has come into our lives in interesting ways. I haven't done any technical work for Los Alamos since I worked on the super, but my time there still has an impact. Rudy Peierls (Sir Rudolf now), whom we met on the Hill, invited us to Oxford, and we spent a delightful sabbatical year there. Mary and the children became villagers, and my fellowship in Timothy Dwight at Yale got me an invitation to participate a bit in college life of Oxford in Queens College.

My research, teaching, and eventually administration kept me too busy to work at Los Alamos during the summers, as Bob Gluckstern sometimes did, but I always kept track of what was going on and regularly sent students there. When we came back to New Mexico—to Mary's especial delight—I was able to bring the university and Los Alamos together for their mutual benefit (Sandia, Kirtland Air Force Base, and White Sands, too) in research and student training. Of course we visited the Hill, and I had the pleasure of standing on a catwalk with Louis Rosen above the "meson factory" and seeing the ports where many of the experiments I had written up for Yale's proposal were set up. Later (1992) I attended and wrote a piece for the proceedings of a conference on "Climate Change and Energy Policy" that Louis had helped put together with international participation. It became possible to bring a university branch at Los Alamos under the guidance of UNM and to tailor programs for Lab employees. I was an unpaid consultant to Los Alamos on proliferation and weapons inspection until I retired from the provostship at UNM. Most recently, I have continued my campaign to inform citizens about nuclear weapons by talking to tour groups and visiting historical sites at Los Alamos and White Sands with them.

When my son John decided to do a suite of paintings on my experience at S-Site, the director got us permission (with Roger Meade's help) to visit the old main casting building before it fell to the wrecker's ball. The resulting paintings are the only pictures of S-Site during the war—however long after they were painted. John is, of course, the illustrator of this memoir. An exhibition of some of the paintings has helped a number of people who are trying to preserve the few buildings left from wartime at S-Site and elsewhere, led by John Isaacson from Los Alamos, Cindy Kelly of the Atomic Heritage Foundation in Washington, and Ellen Bradbury Reid of Royal Road Tours in Santa Fe.

The government feels no responsibility for preserving these buildings and expects private support to accomplish whatever is done.

From the time I got off the train at Lamy over sixty years ago, it seems, Los Alamos and the work going on there has been a presence in my life (and, willy-nilly, in the lives of Mary, John, and Wendy), even though actual work for the Hill occupied only a few years of my career. Los Alamos is not the intense place it was in 1944–46, but it has become one of the premier scientific laboratories in the world.

In retirement, it is possible to pursue any activity one pleases, including writing memoirs. But it is impossible for a physicist to retire from the study of the subject, I believe, so I keep in touch with what is going on in general, and in particular, learning enough about string theory to understand what younger people are doing. We shall never understand all there is to know about the world and the people in it, but it is fun to try.

Bibliography

I am not a student of things written about the Manhattan Project and the hydrogen bomb, but I have read a few of the books written about them. Family members, knowing my interest, have gotten me some of the books noted here. For readers of this memoir who wish to have a broader picture of the work and some of the people involved, I share my observations of the books I have read. I list them in the order I read them.

■ Smyth, Henry DeWolf, *Atomic Energy for Military Purposes*, Princeton
   University Press, Princeton, 1945.
This is a reprint of what we called "The Smyth Report," published at Los Alamos just after the war ended. It is an excellent, brief review of the history and work in the Manhattan District, starting with the Uranium Committee, of which Breit was a founding member. Breit's effort in organizing the experimental work to measure quantities necessary for the design of a nuclear reactor and bomb is mentioned. He turned over the chair of the Fast Neutron Committee to Oppenheimer in 1942 as the work was reorganized under the Manhattan Project.

■ Lamont, Langston, *Day of Trinity*, Signet, New York, 1965.
This is the first book written by laymen about the project. Lamont interviewed people from Los Alamos, leaders and workers. We appeared together on a panel a couple of years ago, and Langston asked me if he had interviewed me. "No," I said. "No one ever came out to S-Site who didn't need to—I had too many chemical explosives around." He carries the story through Trinity in a fast moving narrative. This book is easier to read than the more technical pieces.

Bibliography

■ Hawkins, David, Edith C. Truslow, and Ralph Carlisle Smith, *Project Y: The Los Alamos Story*, Tomash Publishers, 1983.
This also began as a Los Alamos Laboratory report in 1961. "Project Y" was the army name for the Los Alamos part of the Manhattan Project. I forget who told me that the report was to be published in a history of modern physics series, but I found a copy in the university bookstore at Oregon State. This two-part piece takes the history beyond the war and tells of the work at Los Alamos as it began to mature into the fine laboratory it has become.

■ Schell, Jonathan, *The Fate of the Earth*, Alfred A. Knopf, New York, 1982.
Schell postulates a total nuclear war between the U.S. and the USSR, with the enormous stockpiles of nuclear weapons then available to both countries expended almost at once. He understands that the radioactive fallout from such a war will do more damage to life on earth than the instantaneous damage caused by the explosions. While the war would have taken place in the northern hemisphere, the fallout would eventually pollute the southern hemisphere as it passed through the equatorial doldrums (witness the Ancient Mariner). The lesson of Krakatoa must be remembered. Schell believes the only animal life left would be cockroaches, because they withstand radiation damage best. I calculated the fallout from the stockpiles I knew about in the 1980s and found Schell had been too pessimistic. A few separated bands of humans probably would survive—if they could live off the land, a talent not much cultivated these days. I had been worried about fallout since the Bikini calculations, but had not thought about the thousands of nuclear weapons available in 1982. We had only one or two to think about in 1946. Sagan's calculations on Nuclear Winter also assumes the big stockpiles of those days, and his postulated reduction of agricultural products reduces still further the chance that the survivors of the fallout will last long afterward. Of course things are much better today, with fewer weapons and less hostility between major powers, but we can never relax given the consequences of any level of nuclear war.

■ Szasz, Ferenc, *The Day the Sun Rose Twice*, University of New Mexico Press, Albuquerque, 1984.
This book by my UNM colleague tells the story of the Trinity Test and focuses particularly on the fallout trackers, who were stationed outside the

"reservation" to see whether an unanticipated change of wind would bring radioactive debris toward inhabited areas near the test site. Hirschfelder and Magee were part of that crew.

■ Rhodes, Richard, *The Making of the Atomic Bomb*, Simon and Schuster, New York, 1986.
This is the definitive history of the Manhattan Project. I do not have enough personal knowledge to check every point of the book, but the parts I can check are so carefully written, I have complete confidence in the rest. There is a fair amount of technical material in the book, but I have found no slips to complain of, and the narrative runs so smoothly that those less interested in technical details can get through them safely.

■ Hoddeson, Lillian, Paul W. Henrickson, Roger A. Meade, and Catherine Westfall, *Critical Assembly*, Cambridge University Press, New York, 1993.
This is a technical history of the Lab between 1943 and 1945—the Oppenheimer years. The technical work is well described for laymen without trivializing it. One of the authors, Roger Meade, played host for John and me in our visit to the main casting building at S-Site a dozen years ago and continues to talk to tours of the Lab.

■ Rhodes, Richard, *Dark Sun*, Simon and Schuster, New York, 1995.
As comprehensive as *The Making of the Atomic Bomb*, this book about the hydrogen bomb is as good on the technical side as well. My sometime colleague and mentor for reading Heitler, Klaus Fuchs, gets a lot of play here, because Rhodes tells about the espionage that gave the Russians a head start in fission weapons. The development of their hydrogen bomb, which proceeded almost as quickly as ours although they started working on nuclear weapons after we did, demonstrates my point that any nation with the scientific establishment and industrial base can develop any weapon that any other nation can. Hence my conclusion—which I trust is simply rational and not chauvinistic—that we must maintain a sound research program in weapons development and design so as to be first with any really new device, while at the same time maintaining a policy that we will not use the apocalyptic weapons first and will work with the international community to prevent their further proliferation. I hope that

Rhodes's observation about the decline in war casualties without nuclear weapons continues to hold.

■ Herken, Gregg, *Brotherhood of the Bomb*, Henry Holt, New York, 2002. Herken has written about the relation of three of the major players in the nuclear weapons story, Oppenheimer, Teller, and Lawrence. His claim that Oppie probably had been a member or fellow traveler of the Communist Party has been disputed—but such speculation has been going on since the end of the war, if not before. I claim no close relationship with Oppie, so I have no informed opinion, but from what I know, I'd find it consistent with the flirtation of academics and others in the thirties with Communism. Of course, there's absolutely no suggestion that Oppie ever compromised nuclear secrets with anyone. I was younger than Oppie, but I studied Marx too (and concluded that his position was not liberal, so I lost interest except to observe that the Stalin dictatorship confirmed my analysis). I knew Teller slightly also, and find Herken's portrait consistent with what I know. I'm sorry I never met Ernest Lawrence, one of the giants of experimental physics. We were apparently near each other during the Trinity Test, but it was dark, and I was only there because Kisty thought I deserved to be, whereas Lawrence was one of the VIPs.

None of these books tells the story of S-Site, as I have tried to do at least partially in this memoir. We were too far away from the effort at Townsite, and few of the leaders working there, except Kistiakowsky, knew anything about chemical explosives.

Index

Note: Numbers in bold indicate illustrations.

Accelerators: Cockcroft and Walton and, 117, 127; development of, 118–24; particle, 129, 141–42
Ackerman, Major, 40
Administration: academic, 137–39
Atomic bomb, 12, 153
Atomic energy: experiments with, 38–39; fission and, 36–37
Atomic Museum, Albuquerque, 84
Atoms, 4; charged, 40; evolution concept of, 4–5; structure of, 5, 8, 35, 36

Becquerel, Henri, 35
Bessel functions, 110
Bethe cycle, the, 128
Bethe, Hans: Bikini Tests and, 82; fusion weapons and, 128; Trinity Test and, 65–66
Bikini Tests, 2, 68, 81–84, 107; cloud chamber effect and, 86; planes involved in, 90; radioactivity and, 82, 90–91, 95–101, 142
Binding energies: fusion and, 127–29; neutron, 39; nucleons and, 36–37
Bohr, Niels, 8, 109
Bomb phenomenology, 81
Bradbury, Norris, 74
Breit Group, the, 116–24, 136–37
Breit, Gregory, 140; Cold War and, 132; Los Alamos and, 151; particle accelerators and, 118–24; radio research of, 116; the author and, 2, 33, 94, 112–13, 125–26; work of, 37, 116–17
Breit-Wigner resonant model of nuclear reactions, 74

Calutron, the, 117–18
Camp Seibert, 20–21
Chain reactions: fission and, 42, 43–44, 46
Chemical: bombs, 71; warfare, 20
Chemical Warfare Service, 18
Chernobyl, 144
Chew, Geoffrey, 75

Cloud chamber effect, 86–87, **87**
Cockcroft, John, 117, 127
Cold War, the: security and, 131–32
Compaña hill, 66–67
Compton effect, 98, 100
Compton, Arthur Holly, 98
Computers: IBM, 120–21
Coulomb: barrier, 136; force, 38
Cowan, Clyde, 109
Critical mass: nuclear bomb and, 46–47; nuclear reactors and, 46–47; of plutonium, 49
Curie, Marie, 35–36, 100
Cyclotrons, 117–19

Daghlian, Harry, 106–7
De Broglie, Louis, 109–10
Democritus, 4
Deuterium, 38; fusion energy and, 127–29
Dirac, Paul, 8
Doolittle, Jimmy, 35

Einstein, Albert, 1, 34, 109; curved spacetime and, 7; general theory of relativity, 6–7, 9; Noble Prize and, 11; theory of special relativity, 6
Electrodynamics: classical, 11; quantum, 8–11; quantum mechanics and, 8
Electron-electron: interactions, 122
Elements: classical, 4; heavy, 39; modern, 4; transuranic, 39, 40, 41, 119
Explosive lens casting, 48–69, **53, 55, 59,** **60**; Comp B and, 50, 54, 56, 60; molds for, 50; nuclear explosives and, 64; system for, 57, 58; Trinity Test and, 63–69
Explosive lenses, 27–31, **28, 29, 30, 33,** 48
Explosives, 2, 16–17; castings for, 17, 27, 29–31; Comp B and, 28, 49, 50, 54, 56; lenses for (*see* explosive lenses); nuclear, 46

Fat Man bomb, 48, 58, 71, 82, 83, 84
Fermi, Enrico, 12, 43, 134; Los Alamos and, 43–44, 75; nuclear reactors and, 46; pile experiments and, 44–45; Trinity Test and, 65, 67–68; work of, 38–39, 41, 96
Final Theory, the, 5
Fission, 5; as energy source, 39; bombs, 129–31, 133; chain reactions and, 42,

Index

Index